Natural Resources Development Methodologies

Natural Resources Development Methodologies

Edited by

U. Aswathanarayana
D.Sc., F.N.A., F.A.Sc., F.T.W.A.S., F.G.S. (I)
Mahadevan International Centre for
Water Resources Management
Hyderabad, India

BS Publications
A unit of **BSP Books Pvt., Ltd.**

4-4-309/316, Giriraj Lane, Sultan Bazar,
Hyderabad - 500 095
Phone : 040 - 23445605, 23445688

Published by

4-4-309/316, Giriraj Lane, Sultan Bazar,
Hyderabad - 500 095
Phone : 040 - 23445605, 23445688
e-mail : info@bspbooks.net

ISBN : 978-93-52300-22-8 (HB)

Foreword

The inspiration for this volume is our guru, the Late Professor Calamuru Mahadevan of Andhra University. Visakhapatnam, who was both an eminent field geologist and a visionary academic, around the time when India attained independence. Natural resources development in general and groundwater development in particular, were his passion with which he successfully imbued successive generations of his students.

When his Birth Centenary was celebrated in May, 2001, forty years after his death in harness, over 200 younger colleagues, students and admirers of his gathered to reminisce and ruminate about his legacy.

Thanks to the initiative of Prof. U. Aswathanarayana at the international seminar organized then, it was decided to set up the Mahadevan Memorial Trust and the Mahadevan International Centre for Water Resources Management. Prof. Aswanthanarayana and I have had the privilege of being closely associated with both.

The Govt. of Andhra Pradesh, having recognized the usefulness of our efforts, graciously allowed the Mahadevan Centre to function as a Partner of the MCR Institute of Human Resources Development, Hyderabad.

We have organized an interactive training course on *"Land surface hydrological processes in relation to water resources management"* at the MCR HRD Institute in Dec. 2002. This was followed by a Refresher Course for university teachers, and a Roundtable on *"Natural Resources Management, Geoscience instruction and Jobs"* held in JNTU under the auspices of the Indian Academy of Sciences. We are happy that both these efforts have been welcomed in well recognized international fora of geoscientists.

There is little doubt that any strategy for the protection of the environment, optimal use of natural resources, preparedness for and mitigation of, natural hazards, etc. has to be *knowledge-driven*. The present volume constitutes an attempt to disseminate updated knowledge on the methodologies for the development of natural resources and the changes needed in the pedagogic perspectives in training the future generation for a balanced and sustained development of man in harmony with nature in the context of emerging technologies.

We place this volume as the first flower at the feet of our guru in reverence.

B.K. Rao
Former Secretary to Govt. of India

Preface

Water, Soils and minerals constitute the most significant natural resources of a community. Innovations in technology are generating new jobs in the process of converting into a resource what was yesterday a non-resource *("Resources are not, they become"* – Zimmerman), in developing process and control technologies to minimize wastes, and in recycling wastes. In the case of the developing countries, any pathway for economic progress has to be *knowledge-driven,* and *job-led,* and the technologies of choice have not only to be ecologically sustainable and economically viable, but more importantly people-participatory. The new kinds of jobs (e.g. in poverty alleviation projects via micro-enterprises based on value-added processing of natural resources) have a strong environmental relevance, and lie at the interfaces of several traditional scientific disciplines.

Science alone cannot guarantee good decisions. The purpose of the science input is to help the policy-makers to make informed decisions through understanding of the implications, uncertainties and limitations of the various techno-economic options placed before them. Incorporating scientific information (e.g. water quality) into the administrative and legislative processes is a difficult task. A practical way to address the problem is to translate complex science and technology information in the form of "Best Available Technology" (BAT) and "Best Practicable Environmental Option" (BEPO) modules, incorporating easily-understandable bench-marks.

The volume has arisen from two major scientific activities under taken under the auspices of the Mahadevan International Centre for Water Resources Management, Hyderabad, supplemented with some solicited contributions to make the volume more comprehensive:

(i) An interactive Training course entitled, "*Land surface hydrological processes in relation to water resources management, and extreme weather events* (*droughts and floods*)", held in Hyderabad, India, Dec. 7-17, 2002, for the in-service personnel (hydrogeologists, irrigation engineers and others) of the concerned Departments of the Government of Andhra Pradesh, under partnership with the MCR Human Resources Development Institute of the Govt. of Andhra Pradesh. The high quality of the course led to its being noticed in *Eos* (American Geophysical Union, Washington, D.C.), of Feb. 11, 2003.

(ii) Roundtable discussion on *"Natural Resources Management, Geoscience instruction and Jobs"*, held in the Jawaharlal Nehru Technological University, Hyderabad, in connection with refresher course for earth science teachers (Nov. 3-15, 2003), under the auspices of the Indian Academy of Sciences. The outcome of the Roundtable Discussions has been a strategy of job generation through natural resources management by giving end-use orientation to geoscience instruction. It has been widely publicized and has been well received globally (vide Report in the journal of the International Union of Geological Sciences, *Episodes,* Mar. 2004).

The present broad-spectrum volume integrates the natural resources management strategies arising from the Roundtable Discussion with the water management issues covered in the first training course, thereby enhancing its coverage and usefulness.

The volume has fifteen chapters divided into four Units

Unit I contains an overview chapter dealing with the sustainability of resource management. Unit II constitutes the core of the volume with twelve chapters. It deals with the techno-socio-economic issues involved in water resources management. By virtue of its advantages of repetitive coverage and capability for synoptic overview, satellite remote sensing has emerged as a powerful and cost-effective tool for resources surveys, and inventories, monitoring, systems of optimal use, of water resources, and for integrating the databases related to water and soil resources, ecosystem, etc. Units III and IV deal with soil and mineral resources.

The book carries a perceptive Foreword by my distinguished class-fellow, Dr. B. K Rao, who as Secretary to the Govt. of India, had a long and fruitful association with the natural resources development in India (he also helped me in organizing the courses).

The book will be useful to university students and professionals in the area of water sciences and technologies, earth and environmental sciences and technologies, resource management, agriculture, ecology, civil engineering, geography, sociology, etc.

Hyderabad, India **U. Aswathanarayana**
October, 2005 **Editor**

Contents

Unit - I

Sustainability in Resource Management

Unit - II

Water Resources

Unit - III

Soil Resources

Unit - IV

Mineral Resources

UNIT - I

Sustainability in Resource Management

1

Sustainability in Resource Industries : Implications for Management, Technology and Education

J.S.J. Van Deventer

Faculty of Engineering, The University of Melbourne, Victoria 3010, Australia. Email <jannie@unimelb.edu.au>

Keywords : Sustainability; Global Warming; Education; Recycling; Management; Social Licence

Introduction

Increased public awareness of the need to protect the environment has weighed heavily on the resources industries. Many resource companies today are serious about implementing sustainable practices and are engaging with the community to ensure mutual acceptance of policies. Nevertheless, the mining and mineral industry has been unable to alter the wider perception of it as being a polluter and one, which does not hold environmental initiatives in high regard. Much of this perception is a result of historical baggage, whereby in Europe the image is still strong of cloth-capped coal miners toiling away underground. This misconception of the mining industry does very little to help a modern day mining company to raise additional funds for environmental projects. The resource industry has always been rather unpopular, but in recent times the industry is failing to control the agenda for its sustainable development.

The debate on the environmental impact of mining operations is almost as old as mining itself. It is ironic that Agricola in 1556 already said: "...the strongest argument of the detractors is that the fields are devastated by mining operations, for which reason formerly Italians were warned by law that no one should dig the earth for metals and so injure their very fertile fields ...Thus it is said, it is clear to all that there is greater detriment from mining than the value of the metals which the mining produces."

"Sustainability" has been defined in many ways, but the most popular definition is "Development meeting the needs of the present generation without undermining the capacity of future generations to meet their needs." Needs should be interpreted much wider than just consumption requirements; they also include aesthetics, cultural and political values. The time scale over which sustainability is evaluated is often neglected in debate, especially when the longer timescales of geological and global change are considered.

In this paper sustainability will be discussed in a wider context than just the environment, as it also encompasses markets, technology and people. It will be explained that mineral extraction cannot accept all the blame for environmental damage, however, shifting the blame is ineffective. The poor image of resource companies results from the sins of the past, but today companies realise that they require a "Social Licence" to operate.

How Many Disasters are Natural ?

Figure 1.1 illustrates that there has been a substantial rise in the monetary value of the impact of natural disasters over the last 50 years. A cynical view could be that with modern communication and increasing globalisation such disasters have been better documented and assessed, and in any case, it is inappropriate to draw conclusions about climate related phenomena over such a short timescale. On the other hand, the proponents of human induced global environmental change view this trend as yet further proof of their hypothesis. It could be argued that people are less and less the victims of "acts of God," and more and more the victims of "acts of man." Increased understanding of environmental processes and a growing capacity for early warning should identify threats and risks. Investment now in environmental management should result in savings later, however, this is difficult to argue in any political system. The impact of such "natural" disasters should be viewed in relative terms. The death toll from infectious diseases was 160 times the number killed in 1999 from natural disasters, including earthquakes in Turkey, floods in Venezuela and cyclones in India. 800 million people (14% of world population) are hungry every day, and lack basic literary skills and education (UNEP GEO Report 3). Therefore, the concept of "Sustainable Development" should also be concerned with these social tragedies and not just "natural" disasters.

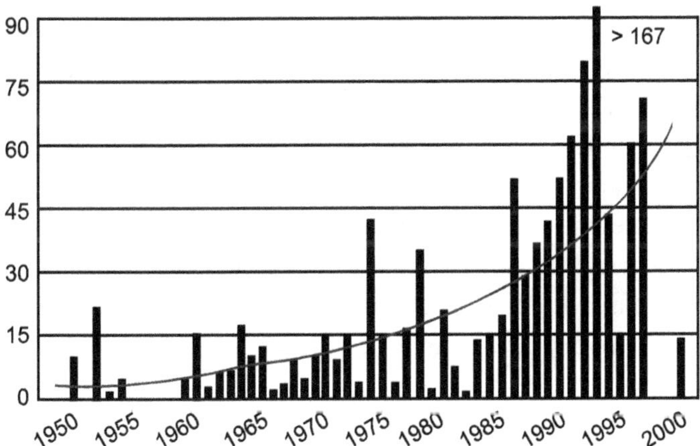

Fig.1.1 Cost of major natural disasters (US$billion).
Source: UNEP GEO Report 3

A Geological Perspective on Global Climate Change

The duration of geological cycles is beyond the comprehension of humans. Therefore, the view of geologists such as Ian Plimer (2003) is that the global warming observed over the last century could be either a "short term" variation on geological scales, or be part of a more long term cycle. At present there are two schools of research, each providing convincing modelling but contradictory evidence about the relationship between global warming and human activity. The skeptical geologists say that the greenhouse propaganda is ignorant of the history of our dynamic planet and focuses on short term phenomena.

For example, 80% of all time planet Earth has been a warm, wet greenhouse. Moreover, polar caps are rare, plants have been here only 10% of the time, grass has been here 2% of the time, 99.9% of all life that has existed is extinct, and O_2 has been in the atmosphere only 50% of the time. Sea levels have risen and fallen up to 400m, land masses constantly rise and fall, thousands of times. Also, global climate changes are not synchronous, so that changes in the northern and southern hemisphere could be different. In any case, Europe was much warmer 900-1300AD. 6,000 years ago North Africa became more desert-like, but the great Mesopotamian cities were established. 5,000 years ago glacier expansion and drought collapsed the Sumerian cities. Geologists like Plimer (2003) suggest that rapid climate change is caused by supernovas, solar flaring, sunspots, meteorites, comets, uplift of mountain ranges, drifting continents, orbital changes, changes in the shape of the earth, changes in ocean currents and volcanic activity, and not necessarily by human generated greenhouse gases. Furthermore, the amount of carbon in the atmosphere is relatively low compared with the quantity in the oceans and lithosphere, as shown in Table 1.1. Therefore, it does not take much of a shift between these reservoirs to make a substantial change in the concentration of CO_2 in the atmosphere.

Table 1.1 Natural reservoirs for carbon dioxide, Gt (Source: H.-P. Ilgner, 2001)

Natural Reservoirs	Global storage (Gt CO_2)
Lithosphere	55,000,000
Dissolved in oceans	140,000
Biosphere	2,200
Earth's atmosphere	2,620

This hypothesis cannot argue that greenhouse gas emissions have not been responsible for climate change. The view that a warmer climate is not necessarily undesirable depends on where you are. Warmer is better only if sufficient water is available. The argument that in the past humans have adapted or moved in response to hostile climate change is not a valid premise for the future. Rising sea levels will flood low countries, and with 8 billion people in 2025 it is impossible to relocate entire populations. The geologist's view is that the universe will find its own equilibrium, so the human induced CO_2 problem is merely a perturbation. In any case, humans will become extinct if we contemplate the chances of any species to survive. The earth will repair our damage eventually, but many of us feel that we do not have that much time! The mineral engineer's view should be: "Watch the kinetics! There is no need to rush it, so let us do something about it."

The minerals industry is directly linked to CO_2 generation and global warming, because it is a significant consumer of the world's total energy supply. Also, the use of coal for power generation is directly linked to mining. Therefore, the CO_2 debate is most relevant to the resource industry.

The Greenhouse Gas Century

Whether one accepts the geologist's view or not, there has been a rapid warming of the earth since the 1970's. The warming of the Antarctic ice-shelves from above and below has resulted in the melting of vast sheets of ice with potentially catastrophic consequences. Recent studies at the Antarctic Institute of Argentina have shown that seas might rise as much as several metres over the next few centuries as a result of these ice-shelves melting. This so-called slow motion disaster, to be borne by future generations, will result in lost shorelines, salinated water supplies, and damaged ecosystems. Lakes forming in the Himalayas and other mountain ranges threaten the lives of tens of thousands of people. These lakes are being created as glaciers retreat and high altitude snowfields melt in response to a 1°C temperature rise over 25 years. Over 44 lakes situated in Nepal and Bhutan are filling so rapidly that they are likely to burst within 5 to 10 years, sending millions of litres of water into deeply populated valleys.

As 85% of global energy use still derives from fossil fuels, this global warming will accelerate as the population rises from the current 6 billion to more than 8 billion by 2025. This population growth requires the equivalent of 1-2 new power stations per week each of 1 GW. The global average per capita per annum energy use is 74 GJ, compared with 15GJ for developing countries, so that rapid economic development will further aggravate the problem. A 50-70% reduction in energy consumption is required to stop global warming (Ilgner, 2001). In the meantime mankind continues talking without much action. The Montreal Protocol on Substances that Deplete the Ozone Layer, 1989, was signed by 182 parties, and is actually operating. The Basel Convention on Minimisation and Control of Trans-Boundary Movement of Hazardous Waste, 1992, was signed by 149 parties, and is restricting the transport and hiding of toxic waste, and to some extent it is inhibiting its generation in the first instance. Earth Summits were held in Stockholm in 1972, Rio de Janeiro in 1992, a "Rio+5" in New York in 1997, and in Johannesburg, 2002. These summits were an opportunity for world leaders to discuss strategies and initiatives that provide for sustainable industrial growth without adverse effects on the environment.

However, it is questionable whether much has been achieved. The UN Millennium Summit in 2000 featured environmental concerns. The UN Framework Convention on Climate Change (UNFCCC), 1994, resulted in the well known Kyoto Protocol, 1997, with 84 parties signed and 46 ratified. This Protocol reflects the reality that most emissions of greenhouse gases are from industrialised states at present, but with both the US and Australia absent, the future of the Kyoto Protocol is uncertain. For this situation to improve, the 21st century should be devoted to action on the greenhouse gas front. Fig. 1.2 shows that Annex 1 countries must reduce their emissions by 13.7% to meet Protocol obligations, while global emissions need to be reduced by 50%.

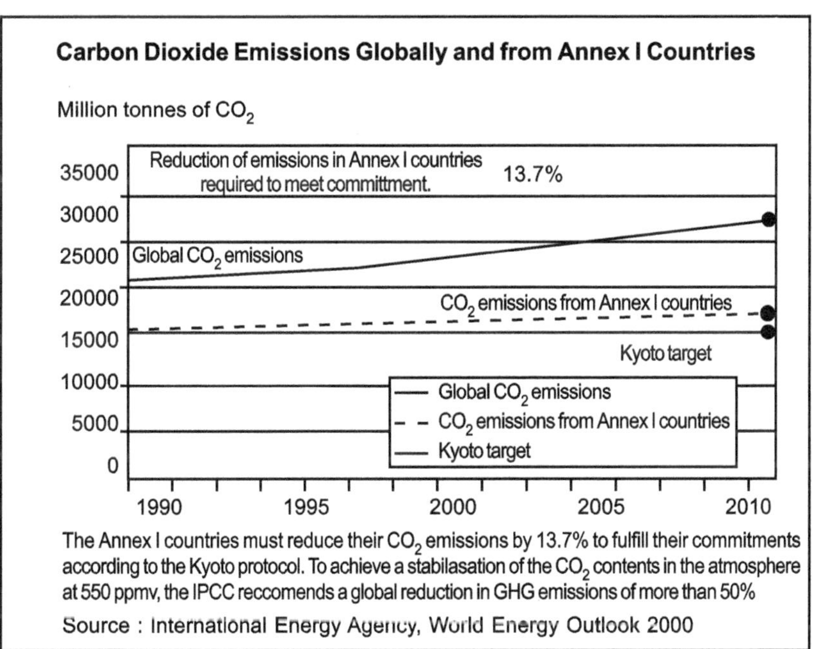

Fig. 1.2 Global CO_2 emissions, including from countries listed in Annex 1 of the Kyoto Protocol, Source : International Energy Agency, World Energy Outlook, 2000.

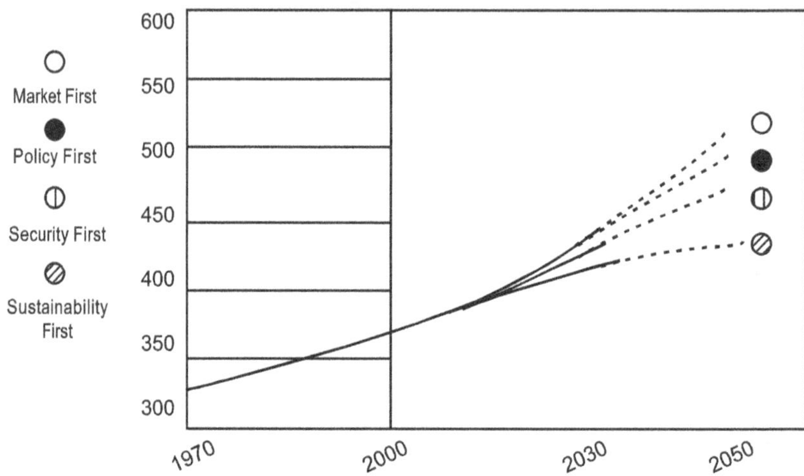

Fig. 1.3 Atmospheric ppm CO_2 for four different scenarios.
Source: UNEP GEO Report 3.

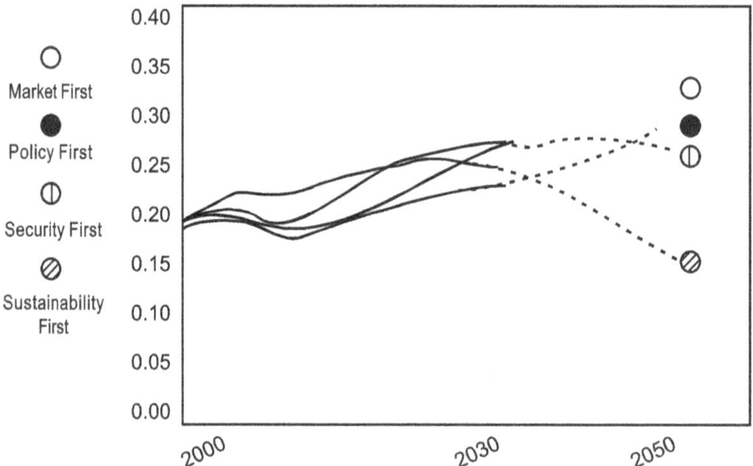

Fig. 1.4 Temperature change (D°C) per 10 years in the atmosphere for four different scenarios. (Environmental damage occurs above 0.1°C/10 years.)
Source : UNEP GEO Report 3

The United Nations Environmental Programme modelled global environmental impact for four scenarios of international strategies, i.e.

1. *Sustainability First* – most positive environmental outlook
2. *Security First* – urban pollution and loss of biodiversity, lack of effective environmental policies.
3. *Markets First* – increased resource demand, gives water shortages.
4. *Policy First* – top down approach

As expected, Figs. 1.3 and 1.4 show that a *Markets First* approach will not deliver a satisfactory outcome, and even a *Security First* or *Policy First* strategy will not reduce CO_2 levels. Only the *Sustainability First* approach, which combines policy with market mechanisms and global agreements, will come close to leveling off CO_2 levels, albeit at higher values than present today, and with a temperature gradient of slightly higher than the critical 0.1°C/ 10 years. Moreover, this modelling has shown that only the latter strategy will result in no increase in the percentage of the population living with water stress.

Shifts in Management Culture

There is a huge gap between how the minerals and resource industries think they ought to be seen and how they are actually regarded by others. Too often loud messages have been ignored about the industry's responsibilities in communities and environments in which it operates. The question is whether public opinion has been shaped by the minerals industry with a sorry track record, and who reacts defensively, or by environmental groups who often distort facts, but whose arguments convince? The way which society at large reacts and whether industry has the perceived credibility is the foundation of the industry's "Social Licence." Today a "Social Licence" is global, not just locally based, so that a company or industry's behaviour in one location may affect its credibility world-wide.

Mineral projects create jobs and prosperity, but are seldom welcomed with opened arms by communities. Telecommunications have brought remote mineral projects to the board rooms and the family lounge. Reputations can be lost quickly, which makes a "Social Licence" imperative.

Most mining companies have now learned to say no thanks to mining in places like Kakadu National Park (Australia) and Kruger Park (South Africa), and not to become entangled in local politics. Failure to gain local (and wider) community support and to be accountable over time will delay a project and give cost blow-out. Companies are also learning to deal with mis- and disinformation, such as the downstream effects of cyanide. Although most companies will minimise waste treatment costs, such expenditure may be justified only against a lack of "Social Licence." It has taken quite a shift in management culture to accept social and environmental responsibility instead of fighting it. "Social Licence" costs are usually up-front in any project, which will stretch weak balance sheets, so that the size of a company is often critical to success.

A number of global collaborative initiatives have been launched by the CEOs of international mining companies. The Global Mining Initiative (GMI) was formed in 1998 and is CEO driven with an aim to turn words into actions. The World Business Council for Sustainable Development (WBCSD) formed the Mining, Minerals and Sustainable Development (MMSD) project in 1999. The International Council on Metals and Mining (ICMM) was formed by the GMI in 2001. It appears that companies now recognise that a change in behaviour and hence management culture is required if the business is to be sustainable.

Full Cost Accounting (FCA), i.e. writing off costs proportionally to disposal rate, is not yet applied to tailings treatment. In general, accounting standards for the extractive industries are not well developed. A methodology is lacking to estimate overall liability, and more input from engineers is required if accountants are to make accounting methods more appropriate for the resource industries. If the financial markets do not understand and correctly value companies that more than others move towards sustainable development, the process of justifying sustainability will be very slow. There is much more work to be done on this subject, even if there are signs that things are beginning to happen. We must be able to argue the business case for sustainable development, as a paradigm based on philanthropy and "do good" arguments will never be acceptable. Efficiency, reputation, and "Social Licence" are effectively all linked to the business case.

Opportunities in Technology

Although alternative energy such as wind farms have been publicised, the public still supports coal-fired power on the basis of low cost. If a country like Australia that is highly dependent on fossil fuels is to reduce its greenhouse gas emissions, capture and sequestration of CO_2 need to be used. However, this technology is still unproven and costly, so the question is will the public and industry pay? This can only happen if the playing field is level globally, and a ratification of the Kyoto Protocol by the USA and Australia is not even close. It will take decades before renewable energy could replace coal capacity, so there is hardly enough time if we are serious about controlling global warming. Coal gasification prior to turbines will reduce greenhouse gases, and eventually a natural gas infrastructure could be replaced by a hydrogen system after 2050. Carbon nanotubes could be used for hydrogen storage., but this is at least 20 years away. Australia has stable saline aquifers near major CO_2 emitters to be used for geo-sequestration, and this is a possibility that needs to be developed. Hence, there are a variety of technologies which the resource industries could develop in conjunction with other industries in order to secure a sustainable future. Proponents of nuclear energy state that it is the only alternative, but that is widely rejected by the community so that a "Social Licence" does not exist. More effective methods of waste nuclear fuel handling and disposal may change this view.

Geopolymers are environmentally friendly materials that could be synthesised from waste materials such as fly ash and slag, resulting in substantial savings in greenhouse gas emissions compared with conventional cement. Research into these materials will promote their commercial acceptance, and hence contribute to a more sustainable resource industry. Reuter et al. (2003) explain that society requires a maximum of end-of-life products to find their way back into the industrial and consumer cycle while minimising both energy input and the impact to the environment, taking cognisance of economics. This task requires that new and innovative designs and production methods must be developed. Hence, the optimisation of the material cycle in a world in which products change rapidly is only possible if the interaction between all technological aspects of creating/using/discarding/recycling products is considered in relationship to fundamental studies including environmental control and policy. It is therefore imperative to achieve sustainability change at various system levels, from global material cycles down to plant and process equipment design and operation.

Implications for Education

For a discipline to flourish it must attract the best young people. It is questionable whether in most countries the resource sector is attracting top talent. In general the image of the industry is still poor; students do not see a great future in minerals and do not believe that mineral extraction is conducted in a sustainable manner. Moreover, there are so many competing and more attractive careers in finance, business consulting, biotechnology and information technology that attract young talent. Young people want to feel they are doing something for society, and in this way the minerals industry has to first obtain a "Social Licence" before it will be embraced.

Even in the days of Agricola (1556) it was acknowledged that minerals engineers should be multi-skilled and have a broad education: "... there are many arts and sciences of which a miner should not be ignorant. First there is Philosophy ... Medicine ... Astronomy ... Surveying ...Arithmetical Science ... Architecture ... Drawing ... Lastly, there is the Law ..." Today students also demand a broad education, and engineering or science is viewed as too narrow. At the University of Melbourne 70% of students study combined degrees, often combining engineering with commerce, arts, languages or law. Students do not want to commit to one industry, and are sensitive to the reputation of a company or industry. Sustainability in the resource industry also depends on recruiting more females into engineering and the minerals industry.

Final Remarks

In general, the resource industries are on the right track in terms of community involvement, emphasis on sustainability and importance of "Social Licence." However, the industry should consider the full materials cycle, not just technically but market-wise. Only in this way will profitability be optimised in a sustainable manner. Whether global warming is caused by human activity or not, it is imperative that the resource industries contribute to the debate and the reduction in emissions. Furthermore, for an industry to flourish it is imperative that the best young people are recruited and retained. It is questionable whether the minerals industry is succeeding in this goal, and herein lies an educational opportunity if approached correctly.

References

Agricola, Georgius, De Re Metallica. Translated from the first Latin Edition of 1556 by H.C. Hoover and L.H. Hoover, Dover Publications Inc., New York, 1950.

Ilgner, H.-P., Carbon dioxide emissions – The challenge for the 21st century, 6th World Chemical Engineering Congress, Melbourne, 2001

Plimer, I., A short history of planet earth, ABC Books, Sydney, Australia, 2003

Reuter, M.A., Boin, U.M.J., Van Schaik, A., Verhoef, E.V. and Eksteen, J.J.,

Pyrometallurgy – The key to sustainable use of materials, XXII International Mineral Processing Congress, Cape Town, 29 September – 3 October 2003.

World Business Council for Sustainable Development (WBCSD), Mining & Minerals <http://www.wbcsd.ch>

Global Mining Initiative (GMI) <http://www.globalmining.com>

Global Environmental Outlook 3 (GEO3), UNEP. <http://www.grida.no/geo/geo3/english>

United Nations Environment Programme (UNEP) <http://www.unep.org>

UNIT - II

Water Resources

2

Management of Groundwater Resources

G. Krishna Rao

Department of Geology, Andhra University, Visakhapatnam - 530 003, India.
(e-mail : audsi2000@yahoo.co.in)

Key words : Aquifer systems, Lithology, Stratigraphic units, Water balance, Groundwater recharge, Groundwater Quality, Andhra Pradesh.

Introduction

The paper elucidates the ways and means of managing the groundwater resources, using the data for the state of Andhra Pradesh as a case.

Groundwater accounts for about 40 to 50 percent of the global water requirement for all purposes put together. Its development has increased by leaps and bounds during the last three decades. In some countries like India, development has been taking place with little concern for the sustainability of the resource. As a result, over exploitation is taking place causing declining of water levels and sea water intrusion in different parts.

Therefore, the concept of sustainable development has picked up momentum globally. Different management methods are being suggested varying from draft control, change of irrigation practices and crop pattern to artificial recharge of the groundwater. Artificial recharge methods range from macro and micro – watershed development to localized roof – water (rainwater) harvesting structures.

Groundwater hydrology is an interdisciplinary subject. Therefore, all the concerned disciplines including hydrogeologists, hydrologists (civil engineers), hydrogeophysicists and hydrometeorologists come out with different concepts of management. It is observed that some of these professionals at the helm of affairs try to adopt some methods designed from their own angle or simply by copying from elsewhere. Many such methods failed to yield desired results. It is found that these methods are applied commonly without proper understanding of the hydrogeological condition and aquifer system in an area. The views expressed here on the management of groundwater resource (by defining different types of aquifer systems and appropriate methodologies for conducting artificial recharge successfully to achieve the desired result) are based on the field experience of the author in different parts of India and Nepal. Results of his field and research studies are given as case studies for different settings. Therefore, they are applicable to similar hydrogeological settings, in general, in any part of the world.

Groundwater, which occurs in the subsurface strata, is controlled in its occurrence and movement by the geological framework of the aquifers. Groundwater recharge is controlled by the soil zone and unsaturated zone of the water bearing formations. The situation is different for unconfined, semi confined and confined aquifer systems. The aquifer setting differs for crystalline and sedimentary formations. Therefore, a thorough understanding of the aquifer system is essential when assessing the groundwater potential and adopting management techniques including artificial recharge. Case studies for different aquifer systems are taken from Andhra Pradesh State, India

Geologically, the state is underlain by hard rocks in more than three - fourth of its geographical area (Fig. 2.1). To a larger extent, the state is underlain by Archaean gneisses and other unclassified Crystallines and Eastern Ghat complex and the Dharwars. Cuddapah and Kurnool formations belonging to upper proterozoic age represent highly indurated sedimentary formations. They behave as hard rocks from groundwater point of view. Gondwana sedimentary formations of upper carboniferous to Lower cretaceous age occupy the Godavary valley and scattered along the coastal plain while the Holocene alluvium is confined to the coastal plain, and delta regions.

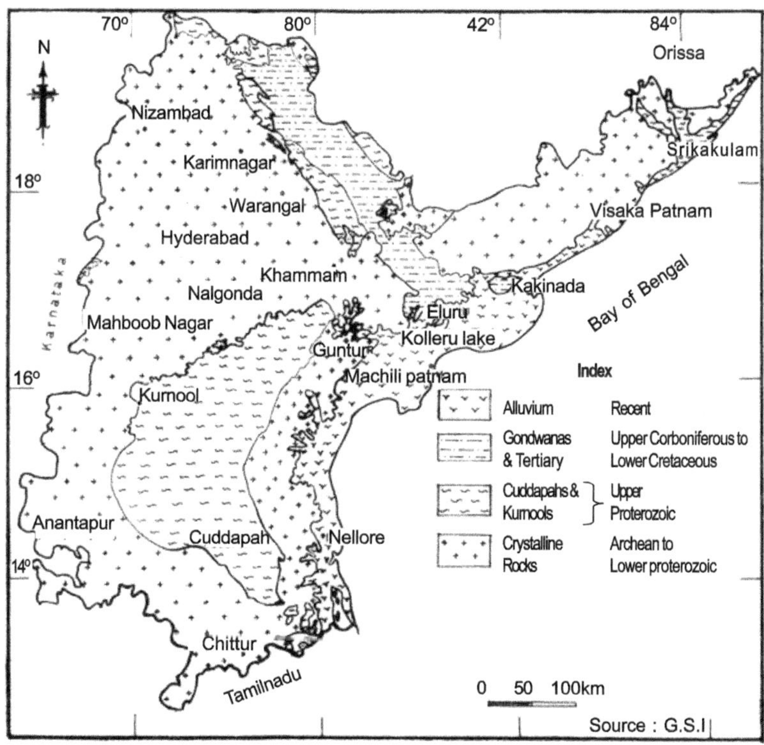

Fig. 2.1 Geological Map of Andhra Pradesh

The Central Ground Water Board (CGWB, 1982) published Hydrogeological atlas for Andhra Pradesh state incorporating all the scientific data. The only change since 1982 would be with respect to draft and consequential fall in water levels. The A.P. Water statistics are given in the Strategy paper on '*Neeru-Meeru*' (Water and You) in February, 2001. The average annual rainfall of the state is 900mm.

Aquifer systems in relation to lithology

Aquifer Systems in crystalline rocks

Groundwater exploration and management in hard rock aquifers, particularly in crystalline rocks, are highly complex because of their heterogeneity, anisotropism and low permeability due to variability in the nature and thickness of weathered zone and fracture zones. They are considered to represent aquitard – aquifer setting. Detailed studies through Canadian Assisted Project in Andhra Pradesh and Vedavati River Basin Project in Karnataka and a few others in other states by the Central Ground Water Board, Govt. of India could unravel the intricacies in these hard rock aquifers. Statistical analysis of bore well data (Subba Rao and Krishna Rao, 1984) and chemical facies studies of dug well and bore well waters located within 30m distance (Table 2.1) in different parts of north coastal Andhra confirmed this setting (Fig. 2.2) and led to doubting even the hydraulic continuity between weathered zone and fracture zones in some parts. It is found that they act as a two-tier system in many places. Isotope studies in the National Geophysical Research Institute (NGRI.), Hyderabad, also revealed that groundwater recharge in to the fracture zones takes place directly from outcrop areas and not necessarily always through the overlying weathered zones.

Table 2.1 Chemical analyses of dug well and bore well waters, Vizianagaram District, A.P.

Sl. no.	Village Name	Sample No.	Ph	T.d.s. Mg/L	Facies
1.	Uttaravalli	03 D 02 B	7.7 7.7	960 1024	Mg. Cl Mg. Cl
2.	Somalingapuram	05 D 04 B	7.9 7.3	556 2034	Mg. Cl Mg. Cl
3.	Garbham	10 D 08 D	7.3 6.9	2084 1024	Na HCO_3 Na cl
4.	Bhyripuram	12 D 09 B	7.8 7.7	564 768	Ca Cl Ca Cl
5.	Chellapuram	16 D 12 B	7.1 6.9	358 512	Ca Cl Ca Cl
6.	Geddapuvalasa	21 D 17 B	7.6 7.3	410 1344	Na HCO_3 Na cl
7.	Rajam	22 D 18 B	8.2 7.4	256 960	Ca HCO_3 Ca Cl
8.	Boddam	23 D 20 D	7.7 7.8	1856 704	Mg. Cl Mg. Cl
9.	Billalavalasa	26 D 24 B	7.8 7.4	768 2304	Mg. Cl Ca Cl
10.	Gotlam	39 D 38 B	7.7 7.9	640 1152	Ca Cl Na.Cl
11.	Ambativalasa	40 D 39 B	7.6 8.3	1568 2368	Ca Cl Ca Cl
12.	Bondapalli	41 D 40 B	8.1 7.6	1248 1408	Na HCO_3 Mg Cl
13.	Bodasingipeta	43 D 42 B	7.8 7.3	3136 5440	Ca Cl Ca Cl
14.	Marupalli	44 D 43 B	7.2 7.8	704 1344	Ca HCO_3 Ca Cl
15.	Marupalli	46 D 45 B	8.1 8.4	1344 704	Ca Cl Ca HCO_3
16.	Arikitota Junction	49 D 48 B	6.8 7.2	304 128	Na Cl Na HCO_3
17.	Burjavalasa	50 D 49 B	7.3 7.7	1024 2624	Na HCO_3 Na Cl

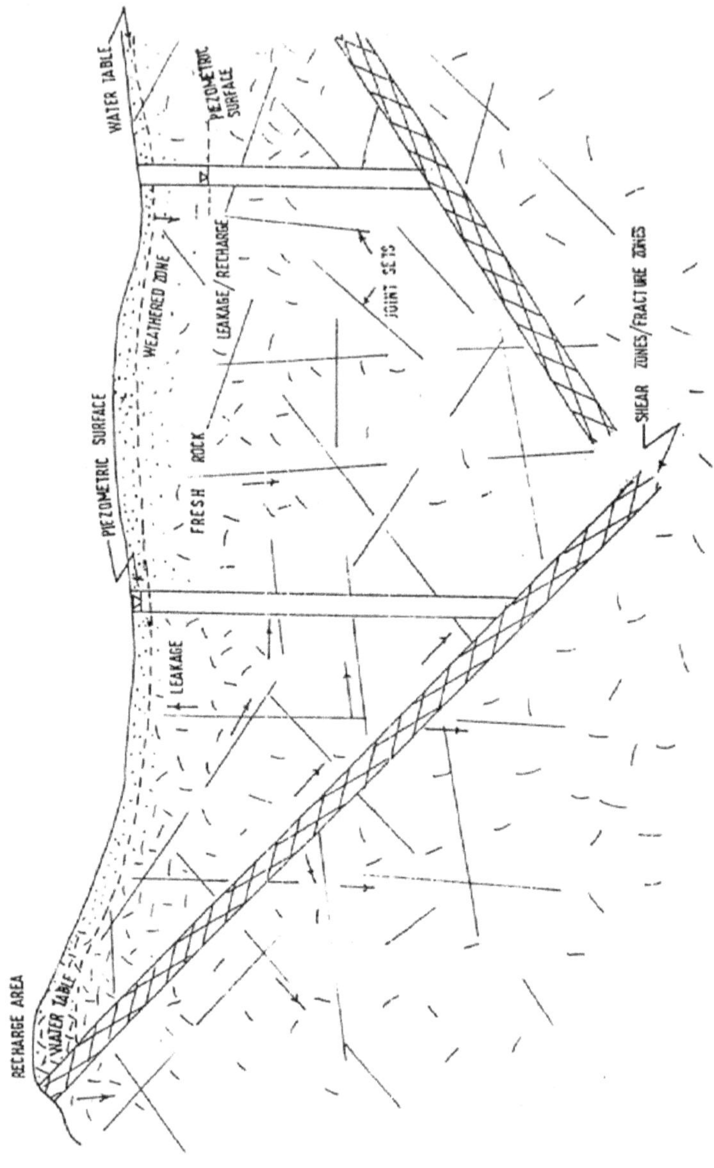

Fig. 2.2 Schematic section of crystalline rock aquifers

Pumping tests on dug wells in crystalline rocks, other hard rocks and indurated sediments revealed that the unit area specific capacity k of wells is, in general, around 1.0 lpm/m^2/m of d.d. and varies up to 5.0 lpm /m^2/ m of d.d (Table 2.2). The flow is non-Darcynian because of very low Raynold number N_R from 8.0 x 10^{-5} to 3.0 x 10^{-3} indicating high friction. Because of very low hydraulic conductivity in these aquifers with K less than 0.1 m / day, the flow is very sluggish at less than about 36 m per year even at a hydraulic gradient of 1.0. Because of this reason, the well performance in shallow aquifer in crystalline rocks and other hard rocks may be considered as highly localized. It is possible that the water table which declined by 20.0 m at a well or a cluster of wells in an area because of heavy draft may remain unaffected in adjacent area about 100 m away where groundwater is not exploited.

Therefore, it is suggested that in times of drought, groundwater exploration may be carried out in virgin areas particularly in valleys, topographic depressions and plains and the groundwater so exploited may be transported to the areas of need. This exercise may be initially restricted for drinking water needs.

Aquifer Systems in Sedimentary Rocks

Sedimentary formations form more or less uniform and homogeneous system of aquifers. However, variations in grain size with lenses and layers of fine grained sediments i.e. silt and clayey sediments create semi-confined and confined nature of local nature within a major aquifer body. Stratified stratigraphic system with alternate shale and sandstone formations however, give rise to defined multi-aquifer systems. In Andhra Pradesh, the Gondwanas along Godavary valley form a major groundwater province. Their groundwater potential is controlled by degree of compaction and the cementing material.

Table 2.2 Hydrological particulars of the pumping tests

S. No.	Location	Geology	Total Depth of the well M.Bgl	Area of cross section of the well M^2	Draw-down M	Recup-eration in 180 min (Rise in water level) M	Speci-fic cap-acity C LPM/ M of D.D	Unit area spec-ific capa-city 'K' LPM/M^2 of D.D	Reyn-olds num-ber N_R $X10^{-4}$	Fann ing 1000/ N_R $X10^6$
1.	Nawabpet	Weathered Granite	6.8	54.3	2.0	0.5	75.7	1.4	7.0	1.4
2.	Nawabpet	- do -	6.0	66.8	2.1	1.1	306.0	4.6	30.0	0.3
3.	Kurnal	- do -	5.8	31.5	1.2	0.2	30.8	1.0	7.0	1.4
4a.	Arakeri	- do -	6.2	59.4	1.2	0.2	75.9	1.3	8.0	1.3
4b.	Arakeri	- do -	6.2	59.4	2.7	0.5	62.3	1.1	5.0	2.0
5.	Bommanahalli	- do -	5.7	72.3	1.2	0.2	78.4	1.1	7.0	1.4
6.	Hallisagar	- do -	4.0	132.0	0.6	0.1	150.5	1.1	4.0	2.5
7.	Gugi	- do -	6.6	42.0	1.8	0.5	73.1	1.7	16.0	0.6
8.	Hunasagi	- do -	4.9	19.6	0.4	0.3	224.0	11.4	4.0	-
9.	Hunasagi	- do -	5.9	29.0	0.2	0.1	111.8	3.9	6.0	1.7
10.	Waginegera	Weathered Kaolinitic	4.6	63.5	0.8	0.2	87.3	1.4	1.0	10.0

Table 2.2 Contd....

11.	Nalwar	- do -	12.0	28.3	4.1	0.8	40.2	1.4	5.0	2.0
12.	Kembhavi	Fractured Granite	10.3	12.0	0.4	0.4/hr	667.2 (720.0)*	55.6	11.0	0.9
13.	Naganur	Chlorite Schist	11.0	90.0	0.9	0.2	127.5	1.84	4.0	2.5
14a.	Shahpur	Chlorite Weathered	15.5	66.0	2.3	0.5	79.1	1.2	7.0	1.4
14b.	Shahpur	Chlorite Weathered	15.5	66.0	1.2	0.4	141.0	2.1	6.0	1.7
15.	Bommanahalli	Basalt	7.1	112.0	0.6	0.2	301.0	1.8	6.0	1.7
16.	Kalhangargi	- do -	5.3	85.5	1.5	0.2	81.0	1.0	9.0	1.1
17.	Hirehangargi	- do -	7.3	60.0	1.5	0.2	31.7	0.5	-	-
18.	Chinamgeri	- do -	9.5	17.5	0.3	0.1	63.8	3.7	1.0	10.0
19a.	Yedrami	- do - (Fractured)	7.4	9.7	0.1	0.1/35 min	898.7 (847.7)*	71.8	5.0	-
19b.	Yedrami	- do - (Fractured)	7.4	9.7	0.3	0.2/35 min	526.5 (722.1)*	54.1	10.0	-
20.	Balbatti	Red Bole Bed	4.5	64.0	1.0	0.2	96.0	1.5	10.0	1.0
21a.	Balbatti	- do -	17.3	3.9	1.7	1.3	30.8	7.8	2.0	5.0
21b.	Balbatti	- do -	17.3	3.9	4.7	2.2	14.0	3.5	4.0	2.5

Table 2.2 Contd....

22.	Limestone	Limestone	7.0	48.1	1.5	0.1	20.5	0.4	2.0	5.0
23.	Madnur	- do - (Fractured, Seepage Zone)	6.7	182.0	0.4	0.2	826.0	4.5	6.0	1.7
24.	Wajjal	(Seepage	2.5	27.4	0.7	0.6/90min	458.1	16.7	60.0	1.7
25.	Chikjewargi	Limestone	9.2	70.6	1.7	0.5	123.0	1.7	9.0	1.1
26.	Andola	Green Shale	9.5	99.0	0.2	0.1	205.9	2.1	0.8	12.5
27.	Jewargi	- do -	12.6	63.6	1.7	0.5	129.5	2.0	5.0	2.0
28.	Bhatpalli	Sandstone (Gondwana)	7.8	17.6	3.3	0.6	21.6	1.2	6.0	1.7
29.	Tundlabuzurg	- do -	8.0	15.4	2.0	0.5	22.8	1.5	6.0	1.7
30.	Tadicherla	- do - (Two horizons)	19.0	31.7	2.5	1.1/90min	190.0	6.0	8.0	1.3
31a.	Kannal	- do -	10.7	46.6	2.6	0.3	31.5	0.7	6.0	1.7
31b.	Kannal	- do -	10.7	46.6	0.9	0.2	66.2	1.4	7.0	1.4

The Gondwanas also occur along the coastal plain, the prominent being in Godavari districts. The Chintalapudi sandstone form a major unit in the type area followed by Tirupati Sandstone of Upper Gondwanas. Raghavapuram shales and Gollapalli sandstone occupy a limited extent. The Tertiary formation, which is called Rajahmundry sandstone, overlies the Deccan Trap basalt. The Rajahmundry sandstone wherever forms an aquifer becomes more potential than the Gondwanas because of its semi-indurated nature.

Since, the yields from the sedimentary aquifers are higher than the crystalline rocks, they are highly exploited for groundwater. Ground water condition in the northwestern part of West Godavary district comprising about six Mandals between Eluru and Chintalapudi is presented here (Fig. 2.3). The Gondwanas form upland between the delta area and foothills of the Eastern Ghats. It was once a backward area because of lack of water sources. Plantations of cashew, mango and rainfed crops were the common sources for living. With the advent of groundwater developmental activity through institutional finance since late seventies, this area turned into a more prosperous zone even envious to the delta people who enjoy canal irrigation. The tube wells which made a mark in eighties with 628 (in 6 Mandals) increased to 18,783 in 2002. The average yields of the tube wells vary from 25 to 38 m^3 /hr in different mandals (counties). Depth of the tube wells range form 60 to 120 m. and in some cases upto 300m. In the crystalline area, the average yield of irrigation bore wells is 11.5 m^3 /hr (Table 2.3). The crops grown in the area are Paddy, sugarcane, palm oil, coconut, Jowar, etc. in addition to plantations, and in many places mixed crops are also grown. With the result, the present draft in the area comes to more than 1100 Million m^3. But the groundwater recharge even with an infiltration factor of 20% is only 173.5 Million m^3, taking into consideration very high infiltration capacity of the sandy formation (overland flow is reported in this area only when the rainfall is 2.5 to 5 cm and more). Such a wide gap between recharge and draft in the area lead to deepening of water table in the unconfined aquifer and all the dug wells to depths of 20m became dry. The farmers have also been deepening the tube wells year by year and increasing the HP of pump sets from time to time. This area becomes a case study of overexploitation of the highest order with draft at 6 times more than the recharge.

Table 2.3 Mandal-wise groundwater draft estimates in west godavari Dt. A.P.

Sl. No.	Mandal	No. of tube wells	Average** yield (m^3/hr)	Average pumping (hrs/yr)	Average draft per year MCM (in 1999)	Average per year MCM (in Sept. 2002)
1.	T. Narasapuram*	3416	25.42	1492	100.58	129.56
2.	Kamavarapu Kota	2660	33.81	1746	115.46	157.02
3.	Lingapalem	3054	11.51	1762	46.90	61.93
4.	Pedavegi	4617	38.34	2062	122.38	365.00
5.	Denduluru*	1703	38.31	2602	114.63	169.79
6.	Dwaraka*	3333	33.34	1960	176.10	217.80
	Toal	**18,783**			**676.05**	**1101.07**

* Small extents in these Mandals lie out of the study area limits

** In 2002, yields decline to 2/3 of 1999

Groundwater Recharge = Geographical Area x 20% of Rainfall
= 937 Km^2 x 20% of 0.95m

= 173.5 M m^3 (after adjusting the crystalline area)

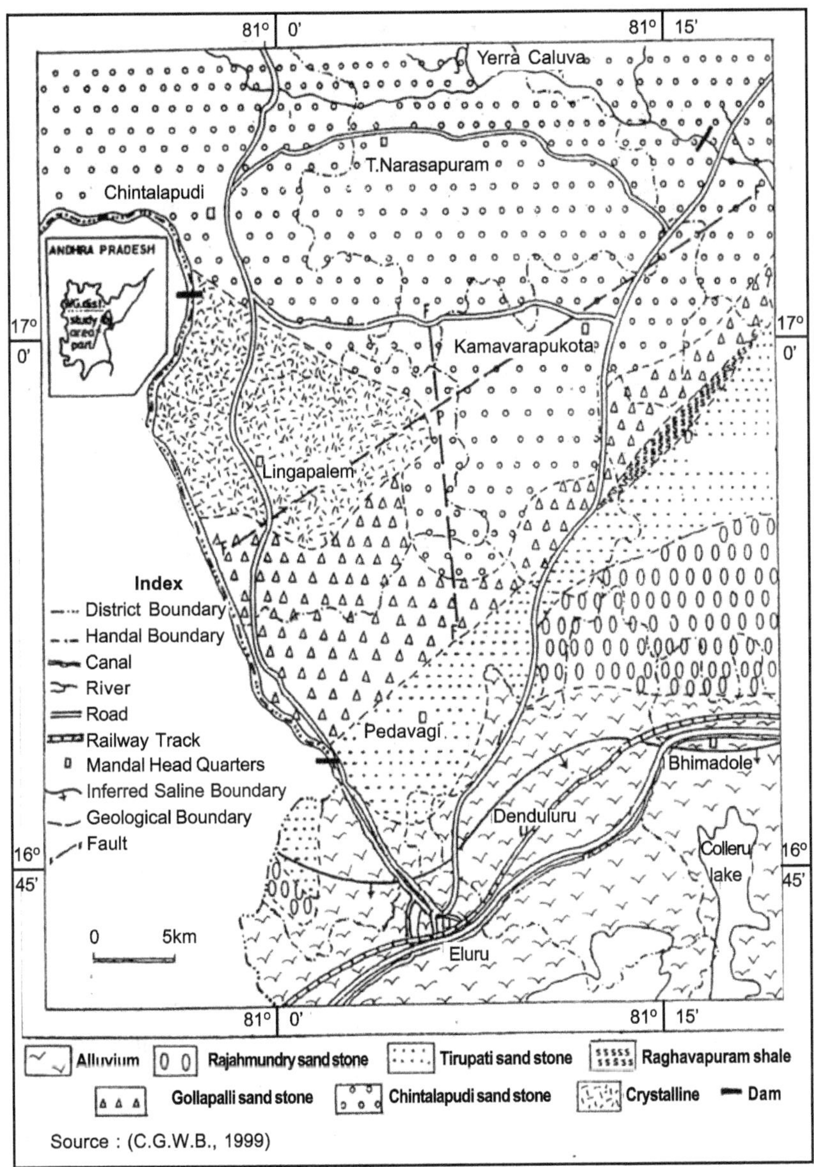

Source : (C.G.W.B., 1999)

Fig. 2.3 Geological map of Western part of West Godavary District, A.P.

Remedial Measures Recommended

The present alarming situation in the study area, can be controlled only by restricting the groundwater draft. This can be accomplished through change of crop-pattern from wet crops to irrigated dry crops and plantations and through proper irrigation practices like drip irrigation and sprinkler systems.

Artificial recharge is also to be practised. However, it has limited scope. It is observed that most of the rainfall is infiltrated in the area as evident by the absence of defined drainage network. A reservoir was built across Errakalava stream on the northern side and a check dam on Tammileru stream on The western side. Excess floodwaters from Tammileru and Gunderu (in the central part) streams are diverted into adjacent chains of tanks thus increasing the water spread. Detailed studies are suggested to understand the subsurface flow system i.e. the paths of infiltrated water. Due to presence of clay layers in the sandstones, the infiltrated waters may be diverted instead of reaching the confined aquifer system. Because of this peculiar situation, artificial recharge has to be taken up mainly through recharges wells. After ascertaining the actual aquifer system in different parts, the appropriate artificial recharge methods may be adopted to regain the original hydrological situation in the area.

Aquifer Systems in Unconsolidated Sediments

Alluvial sediments in the Delta areas and flood plains along major stream courses form highly potential aquifers because of sand and gravel beds which possess maximum hydraulic conductivity. Some prominent areas are Mandapeta area in East Godavari Dt. Kovvuru and Tanuku areas in West Godavari district, Rajampeta, Proddutur, Rayachoti areas in Cuddapah district and Palakonda-Bobbili areas in Srikakulam districts. In delta areas, these zones fall in palaeo-channels of Godavari River. Generally, overexploitation situation is not encountered in these aquifers (as observed in early seventies) since induced recharge takes place during pumping (Table 2.4). However, if the draft becomes more than even the induced recharge, they may undergo declining yields and water levels temporarily during lean period. They will be, however, replenished fully during next flood periods.

Coastal Aquifer Systems

The eastern boundary of Andhra Pradesh state is marked by the Coastline. The coastal plain varies in width from nil as at Visakhapatnam - Bhimunipatnam to as wide as 80 Km in Krishna and Godavary deltas. Many parts in the Coastal Plain suffer with acute drinking water problem because of saline groundwaters. Quality of groundwaters in these areas depends on the geological formations and also sea incursions in the geological past. Crystalline rocks of the Eastern Ghats occur very close to the coastline for over 150km between Tuni and Bhimunipatnam. Sedimentary formations occupy the uplands of East and West Godavary districts and in detached pockets in Srikakulam, Guntur and Prakasam districts. Unconsolidated sediments occur in the deltas of Godavary, Krishna and Pennar rives. Coastal alluvium forms a layer above these formations with varying thickness. Groundwater conditions in different geological terrains along the coast are explained below:

Crystalline Terrain : The crystalline area has topographic advantage with undulating terrain and forming high land near the coast. Therefore, water table/ piezometric surface elevation lies much higher than the mean sea level and hence, gives little scope for seawater intrusion. Even when the water level declined by 10 m below the sea level as at Visakhapatnam beach road, the bore wells are found to yield fresh ground water. This is because of limited hydraulic continuity due to massiveness of the rock and sporadic joints and fractures in them. However with continuous usage, these bore well waters are found to be gradually becoming saline. Total dissolved solids in these bore wells vary from about 600 to 1216 mg/l, the higher towards the coast (Fig. 2.4).

Table 2.4 Hydrological particulars of the pumping tests

Sl. No.	Flood plain of river	Location	Depth water level, m(bgl)	Total depth m(bgl)	Dia-meter of the well (m)	Disch-arge m^3/hr	Draw down (m)	Duration of pumping hr-min	Sp. capacity lpm/m of d.d.	Unit area Sp. capacity city (k) lpm/m^2/m of d.d	Trans-missivity m^2/day
1.	Papaghni	Peddaputta (a)	2.77	6.77	2.20	67.75	1.11	18-55	1017	268	1820
					2.20	126.8	1.90	2-40	1113	293	-
		Cuddapah Mandal (b)	2.98	9.90	3.70	61.2	0.28	2-28	3642	338	-
2.	Penneru	Dorasanipalle Proddutur Mandal	5.03	9.10	2.45	70.8	0.70	26-30	1685	357	1155
3.	Cheyyeru	Nallathimmayyapalli Rajampeta Mandal	2.85	5.65	2.85	69.7	0.53	15-50	2191	344	699
4.	Cheyyeru	Thimmasamudram Rayachoti mandal	7.19	10.20	5.4	55.6	1.78	7-00	520	22.7	594
5.	Cheyyeru	Rayavaram Rayachoti mandal	5.22	6.85	2.1	48.6	1.51	6-05	536	163	764

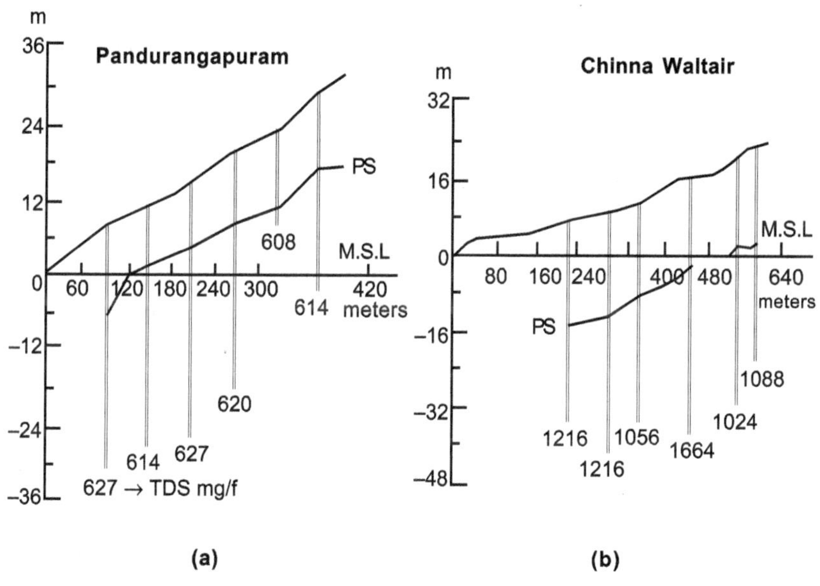

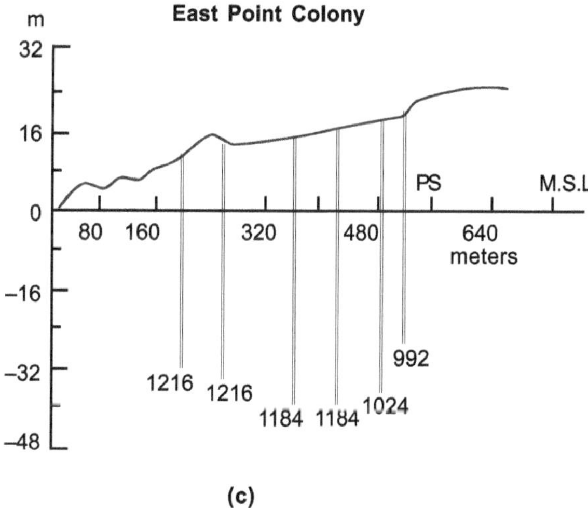

Fig. 2.4 Topography and piezometric surface from shore line to inland

Sedimentary Terrain : In the sedimentary formations of East Godavary district between Tuni and Pithapuram, TDS vary from less than 500 to 6955 mg/l. Best quality of water with less than 500 mg/l occurs in sand dunes along the coastline. In this area, groundwater quality suffers not because of sea water intrusion at depth but due to inundation of the entire area in the geological past during sea transgressions. Even the shallow groundwaters show high salinity. Deep tube wells drilled by the C.G.W.B. at Kommanapalli and Madhavapuram about 7 km west of the coast are found to yield saline waters, but in some other parts the tube wells yield fresh waters. Some lineaments and paleochannels are considered to be responsible for landward intrusion of sea water.

Heavy exploitation of ground water at Chebrolu about 10 km form the coast lead to decline of piezometric surface by about 10m below mean sea level. But still, seawater intrusion could not take place because of confining nature of the aquifers. If this situation continues, gradual salinisation may take place because of leakage of seawater from adjoining saline zones. It is interesting to note that sea transgressions contaminated the aquifers of Gondwanas and even Crystallines in Tuni – Kathipudi area upto the foothills of the Eastern Ghats (Vinoda Rao and Krishna Rao, 1984, Sathibabu et al 2000). The chemical facies and total dissolved solids of ground waters reveal this (Fig. 2.5).

Delta terrain : Delta sediments are formed under changing marine and fluviatile environments. Sea transgressions are recorded with palaeo-strand lines, the farthest being at 30 to 40 km away from the present coastline (Pippara) in West Godavary district (Sambasiva Rao and Vaidyanathan, 1979). Three more strandlines are traced towards coast along Bhimavaram - Palakollu, Kalipatnam - Mogalturu and near Perupalem. The same strand lines extend across the delta towards east up to Uppada near Kakinada. Palaeo-channels of Godavary distributary system are also marked at many places. Prominent are in the areas around Tanuku and Veeravasaram. Kolleru lake which extends upto Eluru marks the Palaeo-marine environment in this part. Therefore, groundwater quality in the delta region is highly variable both laterally and vertically. Fresh groundwaters with 500-992 mg/l occur on some strand lines while the brackish to saline groundwaters contain TDS at 1926 to 14,848 mg/l in the low lying clayey zones (Fig. 2.6). Similar situation is found in other deltas also.

The fresh water zones in the palaeo-channels need to be demarcated and protected from contamination with saline waters from the adjacent or underlying formations through appropriate pumping limits or physical barriers like subsurface dykes.

Groundwater Recharge

A rough estimate of water balance in the state based on the hydrogeological condition is given below:

Evaporation effect on the soil moisture exists up to 0.7 to 0.8 m depth. This soil moisture deficiency is replenished with the first showers of monsoon. Therefore, the maximum absorption of rainfall under this parameter would be equal to the porosity of the soil (if the soil is completely dried), which is about 43% for red sandy soil or alluvium. Therefore, the rainfall that is adjusted into the soil moisture per sq. meter of an area comes to 0.34 m^3/m^2 i.e. 37.7% of the annual rainfall. The transpiration effect varies form place to place depending on the vegetation. Therefore, the figure given by the state Water Statistics at 41% for Evapotranspiration may be taken as a standard value.

Infiltrating rainwater or surface water beyond soil saturation percolates into the deeper layers of the formation and is added to the groundwater. Some part of this water may be lost as interflow through the contact between soil and the underlying formation emerging out as seepage water if there is topographic gradient.

Groundwater recharge to the shallow aquifer may be taken on average as 3 m fluctuation zone though it varies from 1 to 7 m in different areas.

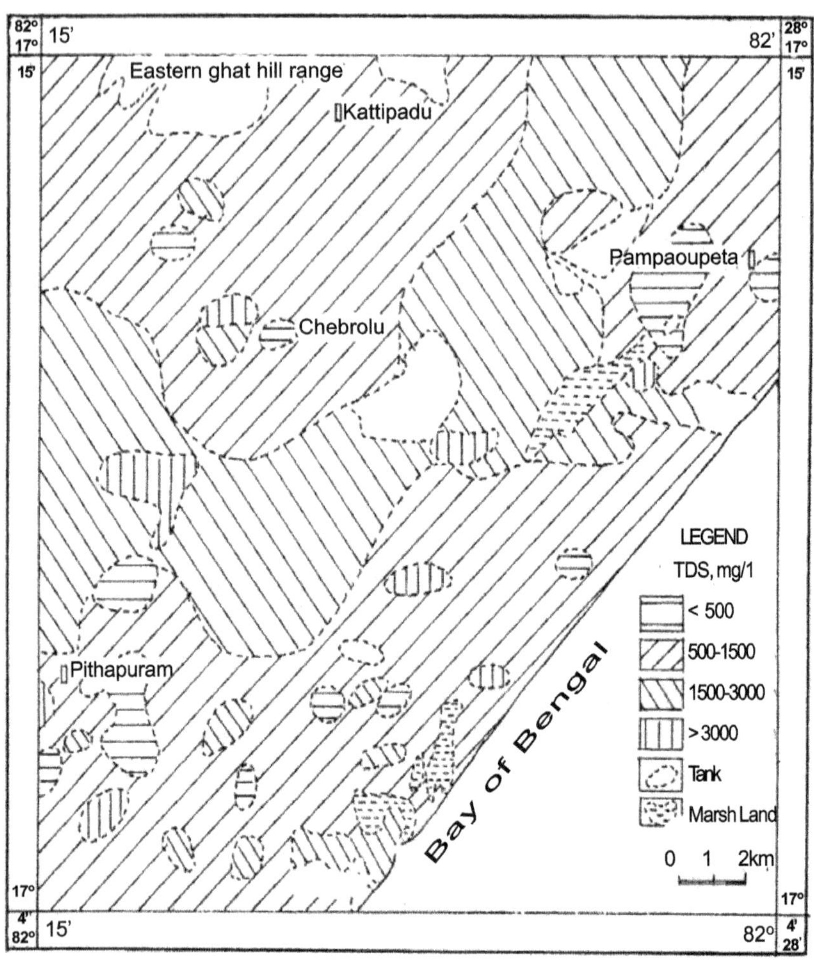

Fig. 2.5 Iso-con map of total dissolved solids in groundwaters of Parts of East Godavary District, A.P (Sathibabu *et al*, 2000)

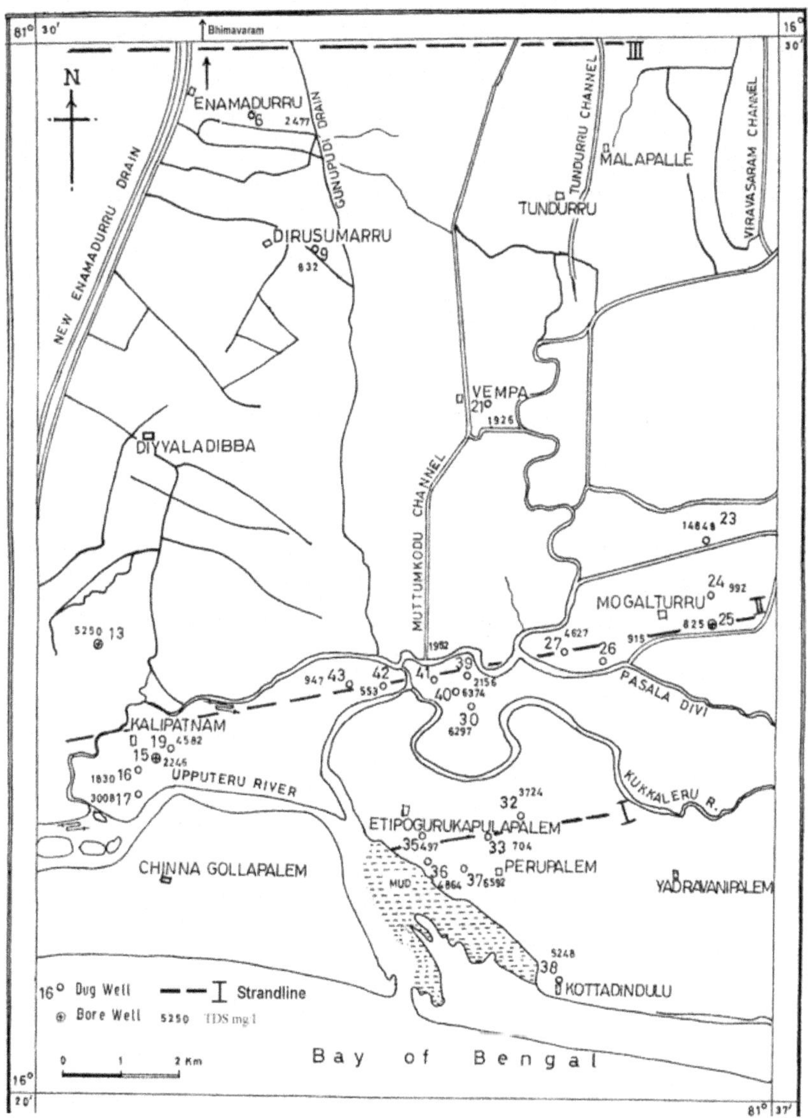

Fig. 2.6 TDS in the groundwaters of southern part of coastal plan W.G.dt. AP.

This component comes to about 0.18 m^3 /m^2 (taking Sp. yield factor as 6% for aquitard type clayey weathered zone materials) which equals to 20% of the rainfall. This figure is variable depending on the nature of aquifer materials and the specific yield factor chosen. The groundwater recharge also takes place to the deeper fracture zones which are under semi-confined to confined conditions, either from the soil zone above or directly through fracture zones from the outcrop areas. Therefore, the 20% recharge factor though seems to be high compared to the 10 % standard figure for hard rocks, may be justified when the total aquifer system in the hard rocks is taken into consideration. The total conservation of rainfall in the form of soil moisture and groundwater recharge comes to about 58% and the balance of 42% goes as runoff. This figure also coincides closely with the state figure of 40% run-off in the state.

Optimisation of Groundwater Recharge : Groundwater occurs in unconfined, semi-confined and confined conditions in the Crystallines and sedimentary formations. Groundwater estimates in the state show that they are calculated, in general, on the basis of unconfined aquifer system. It is perhaps presumed that these aquifers act as source beds for recharging the semi-confined and confined aquifers. But all theses aquifers may be considered as separate units for detailed estimates.

Post-monsoon water table in the month of November (1982) is at less than 3.0 m bgl in the coastal plain and along some parts of the river valleys of Godavary and Krishna (CGWB, 1982). Deep water table at 6 to 12 m bgl is located in some central parts of the state in Telangana and Rayalaseema regions and in the hilly regions of the Eastern Ghats. Interestingly, water level fluctuation shows a minimum of 1 m along the River valleys of Godavary, Krishna and Pennar rivers and also along the Coastal plain. Higher fluctuation up to 7 m is recorded in the central parts of the state. The higher fluctuation coincides with deep water table and minimum fluctuation with the shallow water Table zones. Rainfall distribution and geology have no relation to the water table depth and its fluctuation. It has direct correlation with topography i.e. shallow water table occurs in the valley and plain areas while deep water table occurs in the elevated or high lands. Though the above description gives a regional picture, same phenomenon can be observed in each watershed at micro-level with similar topographic expressions.

The above description reveals that groundwater recharge is maximum on the topographic highs and minimum in the topographic lows irrespective of rainfall and aquifer characteristics. But the real situation is contrary to it i.e. in the coastal plain the rainfall is high, the alluvium possesses high permeability and specific yield compared to the hard rocks in the central parts. This enigmatic situation may be explained as due to the subsurface outflow and inflow conditions in the two topographic divisions and groundwater draft. This situation infers that low recharge in the valleys and plain is because of poor development of groundwater and secondly more subsurface inflow from the adjacent high lands and low rate of outflow. By creating more space in these aquifers through overexploitation of groundwater in the initial stage (taking precaution of hydrodynamics with sea water in areas very close to the coast line) more groundwater recharge can be achieved on par with the high lands. This is called deferred perennial yield. The optimum quantity of draft can be found out at the optimum development stage. This process would harness more rainwater for groundwater recharge in the valleys and coastal areas.

Induced Recharge : Groundwater recharge is also estimated based on the infiltration factor of rainfall at 20% for alluvial areas and 10% for the crystalline areas. But, for delta area like Kovvur, Tanuku, Mandapeta which is underlain by alluvium of the Godavary river, this method may not hold good. A similar situation on the flood plains of Papaghni and Cheyyuru rivers in Cuddapah district proved that the water levels remain stable for continuous pumping of about 24 hours indicating induced recharge from the river beds. In a similar setting in Lumbini zone of the Terai plain in Nepal (semiconfined and confined aquifers), it was proved based on fluid potentials and well hydrographs that the tube wells draw water mostly from the palaeochannels which are connected to the Tinau river (Krishna Rao et al., 1996). In Navalparasi district of the Terai plain also the groundwater recharge has been proved to be (based on Chemical parameters) mainly from the existing river courses and hardly from the rainfall infiltration through the surficial alluvial deposits (Dinesh Pathak and Krishna Rao, 1996). In the present case also, the groundwater recharge should be assessed based on the water table / piezometric surface behaviour and the rate of groundwater flow from the recharge boundary and not on the rainfall infiltration factor.

Watershed Development and Artificial Recharge : Management aspect of groundwater resource was given the highest priority in the very beginning of groundwater development in the country. Groundwater departments were established in early seventies of the last century in all states to implement IDA (USA) schemes. The main reason was to assess well spacing to avoid overexploitation. Since the control was only with respect to wells constructed through financial institutions, it was not effective. As a consequence, overexploitation of groundwater is reported in some areas. It is heartening to know that the area under groundwater irrigation in the state has increased during last one decade form 13.4 to 19.54 lakh hectares while it has decreased from 18.68 to 16.49 lakh hectares under canals and from 9.69 to 7.20 lakh hectares under tanks. Groundwater irrigation forms, therefore, about 47% of the total irrigation in the state.

Water conservation and groundwater recharge through artificial recharge structures like check dams and percolation ponds help to increase the groundwater resource since the impounded water percolates through the substrata for longer time and at higher rate than the rainfall infiltration because of the hydraulic head. However, the areas that require artificial recharge and also suitable for such a measure are to be scientifically selected for fruitful results.

Post –monsoon depth of water level is to be taken as criteria and not the summer level for selecting the groundwater deficit areas. The areas with water table at less than 2.0 m bgl are considered as waterlogged areas and hence not suitable for artificial recharge. Most ideal locations would be at the upper reaches of a drainage basin where the post-monsoon water table is at around 5.0 m bgl. In the hard rock areas, the influence of the recharge ponds may be less than 100 m from the structure since the hydraulic conductivity is very low (<0.1 m/day). A cluster of structures is to be planned to recharge an area of one sq. Kilometre.

It is suggested that the areas with shallow water table conditions may be overexploited initially till the summer water table lies at 8 to 10 m bgl so that an optimum of 5 m or more rise can take place during monsoon period instead of 1 m or less in the present situation. This figure may be modified subsequently based on field observations.

Any additional recharge in the water-logged areas would cause salinisation of soils which ultimately lose their productivity. Many parts in the Indo-Gangetic plain under well irrigation are facing the threat of salinisation of soils.

Depth of Percolation Ponds : Infiltration rates for clay, silt and sand are given as 0.2, 2 and 4 cm / hr respectively. Therefore the intake of rainwater per day is 0.00288 m^3/ m^2/ day in case of clay to 0.057 m^3/m^2/ day in sands. The rainfall should be accordingly 2.8 mm and 57 mm / day. Any excess to these amounts result in overland flow in the clay and sandy area respectively. The rainfall would reach the water table at an average depth of 6.0 m bgl in 6 days in sandy aquifers and 125 days in clayey weathered zone aquifers. Field observation also confirms that the wells in the clayey weathered zones get replenished only by the end of October since the onset of monsoon in June.

The above figures show that infiltration in the hard rock areas is very low at 0.2 cm/hr. Infiltration of water from percolation ponds for about 200 days after monsoon period would be 0.576 m^3/m^2. The percolation pond therefore can hold a water column of 0.57m /m^2 area. But during this period, the evaporation loss would be about 1.0 m (annual evaporation loss in semi arid areas is given as about 1.5 m / year), which means the total water column in the pond should be about 1.57 m. The Infiltration versus evaporation loss is therefore 1:2. Though increase of hydraulic head in the pond would increase infiltration rate, it is not proportional. Increase of area of water spread, on the other hand, has proportional effect on recharge. Therefore, the available surface water may be impounded in more number of ponds thus increasing the area of spread of water. Evaporation losses may be kept at minimum by applying natural or synthetic coverings.

Cost-Benefit Factor : Studies in some southern states reported that about 35% of the percolation tanks in a neighbouring state are totally unsuccessful. Recharge efficiency of percolation tanks in general varies from 20 to 60% depending on the hydrogeological conditions. Evaporation losses directly and through soil moisture come to about 60 %. Secondly, the efficiency of the tanks is reduced in the second year by about 15% (Kulkarni et al., 2001, Muralidharan, 2001). Therefore, the effective life of a percolation tank may be considered as hardly 5 years.

The state government has spent about Rs. 4300 million to create 169.5 M m^3 of storage volume through trenches, percolation ponds, pits etc. Considering the maximum recharge efficiency as 50% of the storage volume of water, cost of the recharged groundwater comes to about Rs. 50/m^3. If the recharge structure is effective on average for about 5 years (though its efficiency decreases year by year), then the cost of recharge water comes to more than Rs. 10/m^3 / year. With the above recharge water, about 17000 ha of paddy may be irrigated at 0.5 ha.m of water / ha. Then the cost of Irrigation comes to more than Rs. 4300 lakhs/ha (one lakh = 100,000) and if distributed over 5 years period it is Rs. 50,000 ha /year for 5 year period without lifting charges. Therefore, these recharge schemes are not economical.

Because of high evaporation loss and low infiltration rates, the percolation tanks may be used for double purpose i.e. for raising horticulture plantations or dry crops apart from groundwater recharge. Irrigation tanks serve this purpose. Tanks with about 1.5 m depth are ideal if it is only for groundwater recharge. In this case, a separate silt-settling tank is desirable. The silt- free water is to be let into the percolation tank for groundwater recharge. This practice would yield better results and increase longevity of the structure. Periodic maintenance of the percolation tank is necessary to keep the infiltration rate constant. This can be done through scarifying or scrapping the bottom surface when dry.

Roof Top Rainwater Harvesting : Rainwater harvesting from roof-tops has been an ancient practice in arid zones. It is a welcome practice in the urban areas where the stress on groundwater has been increasing and recharge area decreasing continuously. But it is suggested that it may be applied in identified towns and identified areas in a town based on hydrogeological situation. The state government has suggested recharge pits for holding 2m^3 of water at a time based on the average roof terrace area and daily rainfall. Dimension of the pit is about 4 times larger than this volume to accommodate a built-in filter bed with gravel and sand. Built-in filter bed has adverse effect on the recharge rate. Therefore, a modified design has been suggested appropriate to the hard rocks (Krishna Rao, 1999). A recharge pit with a trapezoidal cross-section and with sloping walls is recommended (Fig. 2.7). It may be covered with a cement jolly for safety. Appropriate protection may be given to retain the sidewalls. This structure would provide more surface area for infiltration. Filter bed, with coarse sand, where necessary, may be placed adjacent to it. This sand is to be desilted periodically.

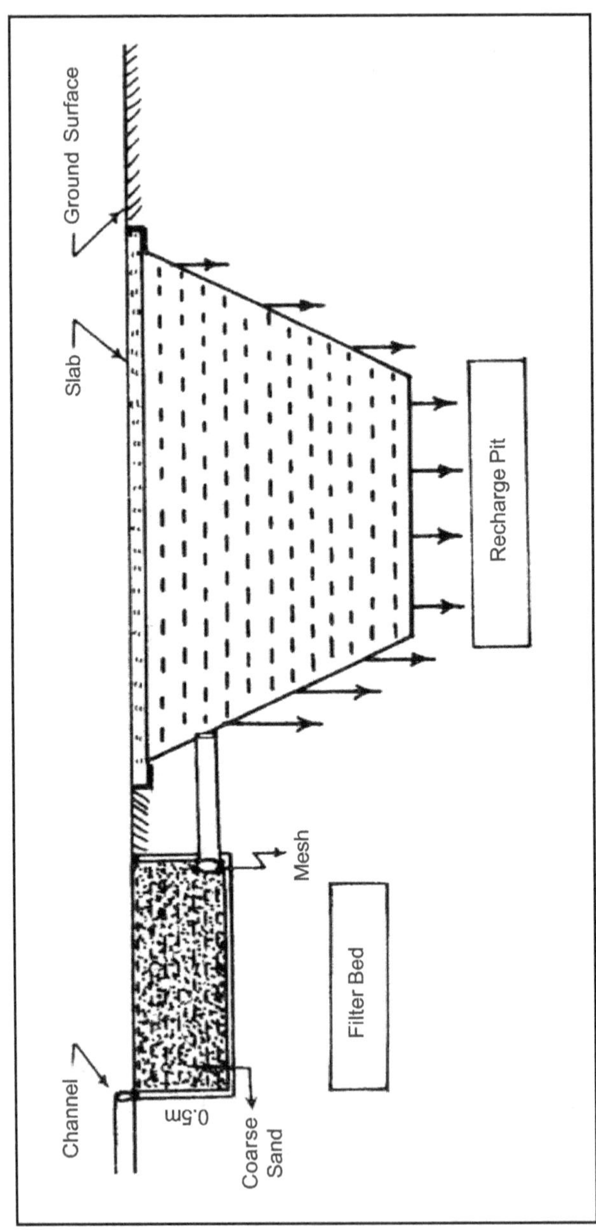

Fig. 2.7 Design of recharge pit

Groundwater Quality

Groundwater is considered as a purified mineral water with many dissolved salts. But certain elements like fluoride, Arsenic, iron in excess amounts would make the water non-potable though the total salinity is within the standard limit. Treatment of theses waters for fluoride and iron are said to be not successful, because they lack palatability. Hydrogeochemistry reveal that these elements may be concentrated in certain geological strata, soil zones and at certain depths (Ramesam and Rajagopalan, 1985, Krishna Rao et al., 1989). In many areas, aquifer zones which are free from these problems may be available at different depths or at different places in the vicinity. Even at the same location, quality of groundwater is observed to vary with depth. Tapping such isolated aquifers may solve the drinking water problems. Problematic waters may be utilised for domestic and other purposes. Where such groundwaters are not available within a reasonable distance, surface water bodies like tanks would be the only alternative. In the coastal areas, dune sands contribute potable groundwaters. These zones have to be protected from local pollution including aquaculture. These waters may be transported to the areas where groundwaters are saline. In some coastal areas with saline waters at shallow depth, Gondwanas and Tertiary sandstones at depth may contain fresh groundwaters and would become good source of potable waters. However their management with appropriate draft is necessary for their sustainable usage.

References

Central Groundwater Board (C.G.W.B.) 1983. *Hydrogeological Atlas of Andhra Pradesh*, C.G.W.B. Publication.

Dinesh Pathak and Krishna Rao, G., 1996. Hydrogeological conditions in the Terai plain of Nawalparasi dt. Lumbini zone, Nepal with special reference to groundwater recharge. *Jour. App. Hydrology*, Vol.9, pp.69-75.

Krishna Rao, G., Bhattarai, T. N. and Sharma, M.P. 1989. A note on the iron-rich groundwaters in the Terai plain of Morang dt. Eastern Nepal, *Bull. Nepal. Geol. Soc.*,Vol.6, pp.13-14.

Krishna Rao, G., Rama Shrestha, Vaidya, Y.L., 1996. Hydrogeological conditions in the Terai plain of Rupandehi dt, Lumbini Zone, Nepal, with special emphasis on groundwater recharge. *Subsurface water Hydrology*, Kluwer Academic Publishers, Netherlands, pp.131-149.

Krishna Rao, G., 1999. A modified design of recharge pit for rainwater harvesting. *Visakha Science Journal*, Vol.2, pp.129-130.

Kulkarni, P.S., Jayant Patil and A.R. Khan, 2001. Role of Geohydrology in artificial recharge. *Proc. Colloquium on basaltic and non-basaltic hard rock hydrology of Peninsular India*, pp.65-74.

Muralidharan, D. Artificial recharge in southern Indian rock systems, *Proc. Colloquium on basaltic and non-basaltic hard rock hydrology of Peninsular India*, pp.103-111.

Ramesam, V., and Rajagopalan, K. 1985. Fluoride ingestion into the natural water of hard rock areas, Peninsular India. *Jour. Geol. Soc. Ind.* Vol.26, pp.125-132.

Sambasiva Rao. K and Vaidyanadhan. R. 1979. Morphology and evolution of Godavary delta, India. *Z. Geograph, N.F*, Vol. 23,no. 3, pp.243-255.

Sathibabu, Y., Subba Rao, N., Krishna Rao, G. 2000. Hydrochemical studies of coastal aquifers in parts of East Godavary District, A.P. *Proc. National Seminar on Groundwater Resources* (ed. Harsh Bhu and V. Agarwal) pp.213-224.

Subba Rao, N. and Krishna Rao, G. 1984. Dug wells and bore wells in hard rock aquifers of Visakhapatnam district, A.P. – A comparative study. *Jour. Assoc. Expl. Geophy*. Vol.4, pp.11-18.

Vinoda Rao, T. and Krishna Rao, G., 1984. Quality of groundwater – A clue to the palaeo-environment, parts of coastal tract between Tandava and Pumpa rivers, E.G. Dt., A.P. *Jour. Ind. Acad. Geol. Sci.,* Vol.26, pp.15-20.

3

Geomorphology of River Basins and Freshwater Aquifers

K.N. Prudhvi Raju

Department of Geography, Banaras Hindu University, Varanasi - 221005.
(rajuknp@banaras.ernet.in)

Key words : Geomorphology, River Basins. Freshwater aquifers. Godavari and Krishna deltas. Sarada River basin.

Introduction

Geomorphology deals with landforms, their morphology, the materials of which they are composed of and the processes (various weathering processes and agents of erosion, transportation and deposition) that shape them. The evolutionary development of landforms are usually related to the bedrock configuration and the deposits which together make up 'the land' or 'landform.'

Vertical and horizontal transmission of surface water through the land and the ultimate storage of the same as ground water depends upon slope, type and structure of the surface and lithology and structure of the underlying deposits and bedrock. Since geomorphology deals with these things in one way or the other, there is a general relationship between ground water (flow systems) and geomorphic manifestations (and geological structure).

For a proper management of a resource and its conservation, the first requirement is a knowledge of the total resource potential. Unfortunately, there is a large gap at this stage itself in our country. And, to have an estimation of total resource potential what is required is a comprehensive understanding of all stages of ground water development in relation to the geomorphic and geologic set-up of the areas.

Though geomorphology is not the ultimate finality in all ground water investigations, it gives a few initial inputs, sometimes, quite significant ones, at least, to plan the 'ground water flow system' investigations and to a certain extent to qualitatively assess the potentials of ground water resources.

The chapter deals with the hydrogeological setting of the freshwater aquifers in Krishna and Godavari deltas, and the geomorphology of the Sarada river basin in relation to its groundwater potential.

Freshwater Aquifers in Krishna and Godavari Deltas

Hydrogeological set-up in the deltas is primarily influenced by depositional history. Three depositional environments (i) the delta plain with its floodplain, meandering distributaries, swamps and beach complex, (ii) the delta front and (iii) the pro-delta-together make up a delta system.

The *delta plain* is made up of marsh deposits and delta front silts and sands. The marsh deposits are covered up by river channel, natural levee and crevasse splay deposits. The deposits of delta plain form (sub-aerial) *topset* deposits.

The *delta front* represents sub-aqueous environment where the distributary channels meet the sea. It is a most active part of deposition (mostly silts and sands) in a delta. Distributary channel extends into the sea with the formation of sub-aqueous levees, distributary mouth bar near the seaward limit of the distributary channel, and distal bar at the seaward sloping margin of delta front. The delta front deposits form (sub-aqueous) *topset* deposits.

Pro-delta is the region seaward of the delta front environment. It is closely associated with prograding delta system. Pro-delta deposits are characterized by fine grained muddy sediments-clay, silty clay. These deposits constitute the *foreset* deposits. The pro-delta deposits are transitional into shelf mud deposits-clay, silty clay and shelly clay and shell layers. These shelf mud deposits are referred to as *bottomset* deposits.

In the present write-up only the aquifers related to the top twenty and odd metres maximum of recent deposits which constitute the modern deltas of Krishna and Godavari rivers (Fig. 3.1) are dealt with. Most of the information given under general aquifer characteristics of Krishna and Godavari deltas is drawn from the paper "Hydrogeological characteristics of deltas of India" by Das, S., published in Memoir No.22 (pp 183-225) of Geological Society of India.

General Aquifer Characteristics in Godavari Delta

The thickness of unconsolidated alluvium in Godavari delta ranges from a few metres at the head of the delta to about 300 m towards the coast. Rajahmundry sandstones lie under this alluvium. The alluvium of modern Godavari delta ranging in thickness from a few to about 30 metres consists of sands, silts and gravels mixed with clay in varying proportions. Ground water occurs under water table and confined conditions. In the upper delta plain (Fig. 3.2), saline water underlies fresh water and in the lower delta plain, with an exception of few selected places, the ground water is saline at all levels.

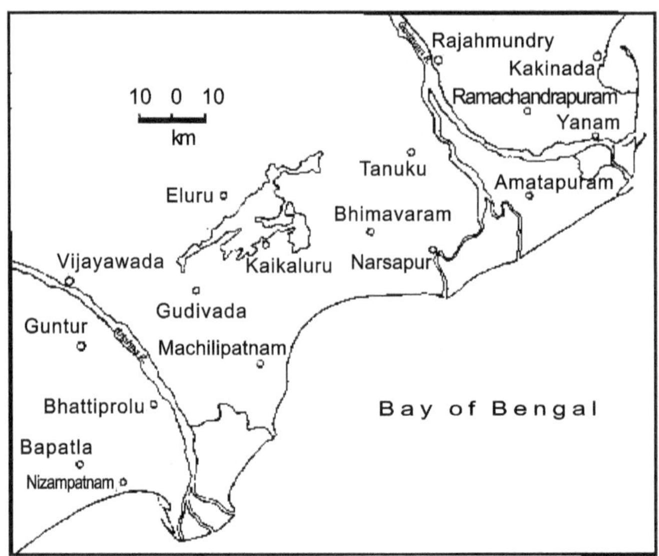

Fig. 3.1 Some selected places in Krishna and Godavari deltas.

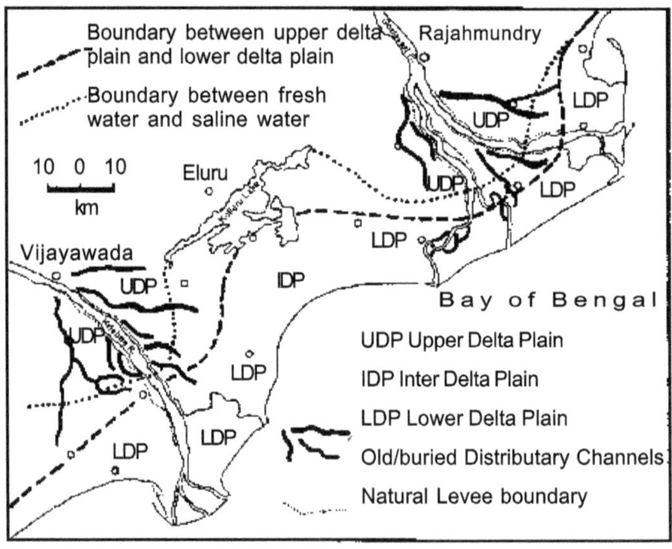

Fig. 3.2 Some geomorphic units, features and ground water zones.

In the northern parts of East Godavari district where there is a good mixture of alluvium from Eleru river, the depth to water table ranges from 2-8m bgl and the yield is quite high from 500-1500 lpm. In the rest of the East Godavari district, the depth to water table ranges from 0.2-8.5 metres bgl and generally it is within 2 m bgl. At depths ranging from 13 to 20m bgl, water yield from confined aquifers is about 4-5 lps (240-300 lpm). In the central delta part of the East Godavari district, the ground water is saline at all levels. But there is good but limited yield to the extent of over 100 lpm from over natural levees, buried channels and point bar deposits.

In the alluvium of West Godavari district, the water table is shallow within 6m bgl. The fresh water is limited to shallow depths locally. The water yield is up to 13 lps (780 lpm). The saline water is encountered below fresh water in the upper delta plain and it is all saline water at all levels in the lower delta plain.

General Aquifer Characteristics in Krishna Delta

In Krishna delta, the fresh water aquifers occur down to a depth of about 20 m. The water table varies from 0.47 to 2.47m bgl. The yield of open wells varies from 10 to 17 lps for 8-10 hours of pumping up to 20 m depth. Tube wells yield about 12 lps on an average.

In the central and western Krishna delta fresh water is generally limited in occurrence to pockets and lenses at shallow depths varying from around a metre to about 14 m bgl. The most promising area for tapping ground water is the northwestern part of the delta close to the river channel. Below 20 m, the fresh water is underlain by saline water in almost all parts of lower delta plain and saline water underlies fresh water in the upper delta plain.

Some Geomorphic Features and Aquifer Characteristics

Buried distributary channels **:** The active distributaries are cut off basically due to channel filling. The channels are slowly alleviated and sealed at both ends isolating the channel loop in the form of a cut-off lake or a cut-off course. In such degrading channels, rate of sedimentation is very slow and some suspended material is introduced during over bank flows. Mainly clayey sediments and organic matter accumulate until filling is complete. Clay and silt is most abundant. Sand is present only in minor amounts. There is coarsening upward sequence in such channels. Ground water occurs in semi-confined condition in such courses and ground water yield is poor in the centres of such channels and good along margins/ levees. The yield in such courses is about 120-150 lpm at very shallow depths of 5-8 m bgl.

Natural Levees **:** Natural embankments bordering the channels, levees gently slope from the river bank into flood plains away from the channel. Natural levees are formed by deposition of sediments when flood waters of a channel overtop its banks. Coarsest sediment is deposited near the channel and grain size decreases away from the channel. The ground water potential is very good yielding up to 200 lpm over the margins of levees at depths of 10-15m bgl.

Flood Basins **:** Elongate areas adjacent to river channels act as settling basins in which suspended fine grained sediment settles down from over bank flows after coarse sediments have been deposited on levees and crevasse splays. The basins represent long-continued accumulation of fine grained suspended sediment. The rate of sedimentation is generally very slow. Usually a silty clay layer 1 or 2 cm thick is deposited during one flood period. In case of actively shifting meandering channels development of thick flood basin deposits is poor and the deposits are found sandwiched between channel deposits. In case of large stable flood basins, backswamps develop in which clay and organic matter dominate. Ground water yield is very poor in such flood basins.

Flood Plains : Rivers that move and shift their position rapidly do not have well developed flood basins. Such rivers develop flood plain deposits with abundant amount of sand. The deposits are similar in nature to natural levee and crevasse splay deposits. And, in such flood plains natural levees are not well developed. The ground water yield in flood plain deposits is better than in flood basin areas.

Crevasse-splay deposits : Usually flooding occurs through concave banks and water occasionally may follow distinct channels cut across the bank through natural levee deposits. The flood water cut channels are called crevasses. Crevasse splay deposits are narrow to broad tongues of sediment. The deposits are coarse grained than natural levee deposits. The orientation of crevasse channels filled with deposits would vary much from the orientation of the main channel. Some of the crevasse splay deposits may continue as thin sandy layers into flood basin facies.

Point Bars : Point bars are depositional features on the concave side of the channel deposited due to channel migration during a flood. They can be as thick as river depth. Coarsest possible sediment available in the river gets deposited as point bars. There is fining upward sequence of deposits. They offer very good aquifer deposits. The yield is 200-300 lpm.

Beach ridges : The beach ridges in the lower delta plain offer local confined aquifers up to a depth of 6 metres maximum. The water is tapped only through open wells for domestic purpose.

Conclusion

Deltas are influenced by a complex combination of fluvial and marine processes. Each delta has more than a dozen distinct environments of deposition and as these depositional environments influence the ground water occurrence and quality, detailed and systematic investigations are necessary before an assessment of ground water potentials is attempted.

Sarada River Basin

The Sarada river in Vasakhapatnam district of Andhra Pradesh with an area of about 2,665 sq. km. flows over a length of about 140 km. and empties its waters into the Bay of Bengal. Topographically, the basin can be divided into high hills, low hills and ridges, and plains. The northern and northwestern part of the basin is of high hills. Down to south, the hills and ridges are scattered here and there all over the plains. The plains are rolling in nature excepting the river built plain. The area comes under semi-arid type of climate and is severely drought prone.

The principal rock units in the area are the khondalite series (quartz-felsfar-garnet-sillimanite gneisses), the charnockite series (hypersthene granites and granulites) and granite gneisses. Structurally the rock formations trend in a general NE-SW direction with the foliation dipping due southeast with deviations in strike towards NNW-SSE, N-S and NNE-SSW, and the folds are predominantly isoclinal.

General Geomorphic Evolution

As for the geomorphic evolution, the basin had undergone erosion and deposition under semi-arid to arid climatic conditions with occasional low-intensity rainfall. The rate of erosion (in the recent past) had been constant giving rise to mostly straight and uniform slopes with pediment surfaces at the foot slopes of hills and ridges. The rolling plain is a resultant of coalescence of pediment surfaces of local extent. The bedrock is shallow (1-23m), uniform and gently slopes towards the drainage lines.

The occasional stone lines in the deposits and the shape of the filling units/sediments conforming to the shape of the channels indicate semi-arid to arid conditions. Also, the channels were shallow in depth and small in width. The climate had changed to semi-arid with intense seasonal rainfall very recently. The cut and fill features and intense dissection of terrace and foot slope deposits, sometimes across the grain of the slope, indicate the change from rainless to rainy semi-arid climate.

The Geomorphic Map

The geomorphic map (Fig. 3.3) presented here gives information on landforms and other geomorphic units under a broad class of exogenous landforms. Three main groups are distinguished, namely denudational, fluvial and maritime landforms. Further, the landforms are classified under two subdivisions in each group viz., destructional landforms and constructional landforms. The landforms, units and features which a relevance to the present study are: pediments, abandoned (buried) channels, river terraces, (creep built) colluvial plain, (creep built) sandy plain, wash plain and river built plain.

Geomorphic Units and Landforms and Hydrological Characteristics

The ground water investigations are limited to only the 'plains' part of the river basin i.e., for about 1600 sq.km. The investigations were carried out by means of a combination of remote sensing and field methods. Fig. 3.4 gives in qualitative terms the ground water potentials of different geomorphic units.

Colluvial Plain

1. ***Location :*** footslopes; occurs immediately after the steep debris slopes of hills and ridges; it is a ubiquitous unit all along and around hills and ridges.
2. ***Slope :*** 3-12 degrees; in granitic and charnockitic terrain slope is gentle 3-5 degrees.
3. ***Extension :*** narrow in width – from a few metres to a maximum of 1 km.
4. ***Material :*** unassorted colluvium consisting of cobbles, pebbles, gravels and an occasional boulder (especially at the bottom) in a matrix of red sand and grit.

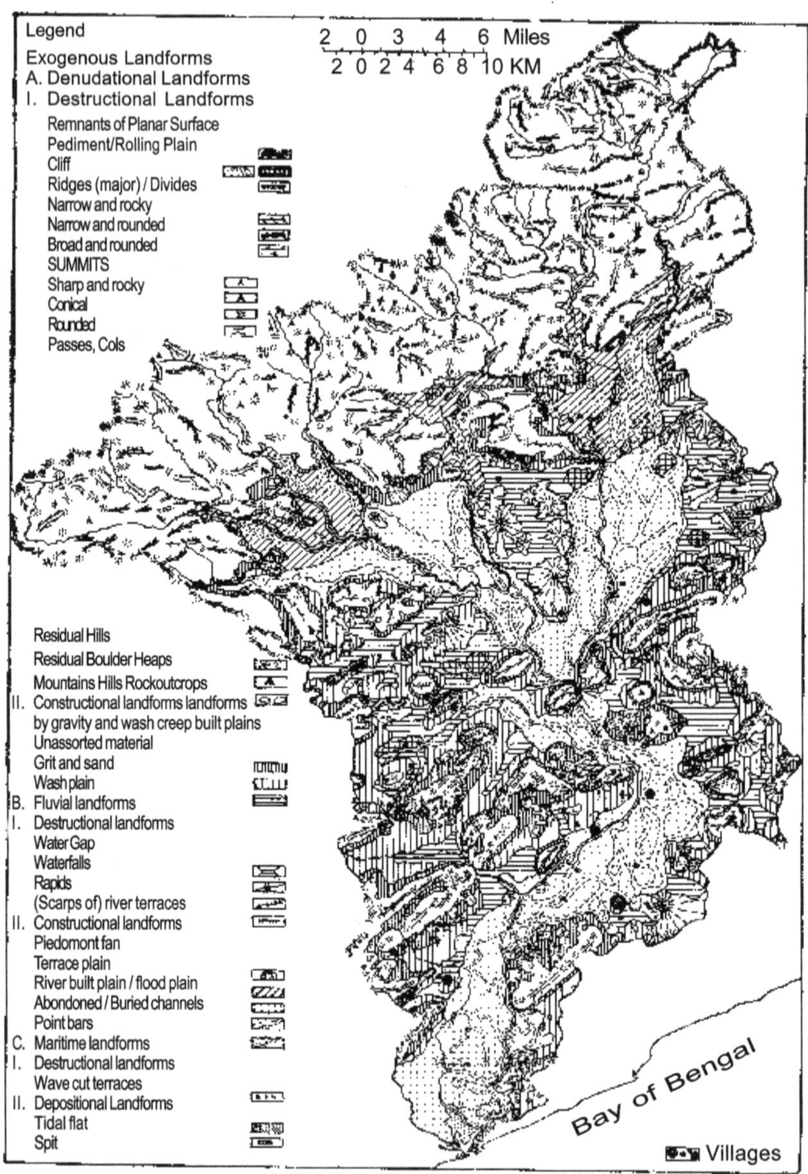

Fig. 3.3 Geomorphic map of sarada river basin.

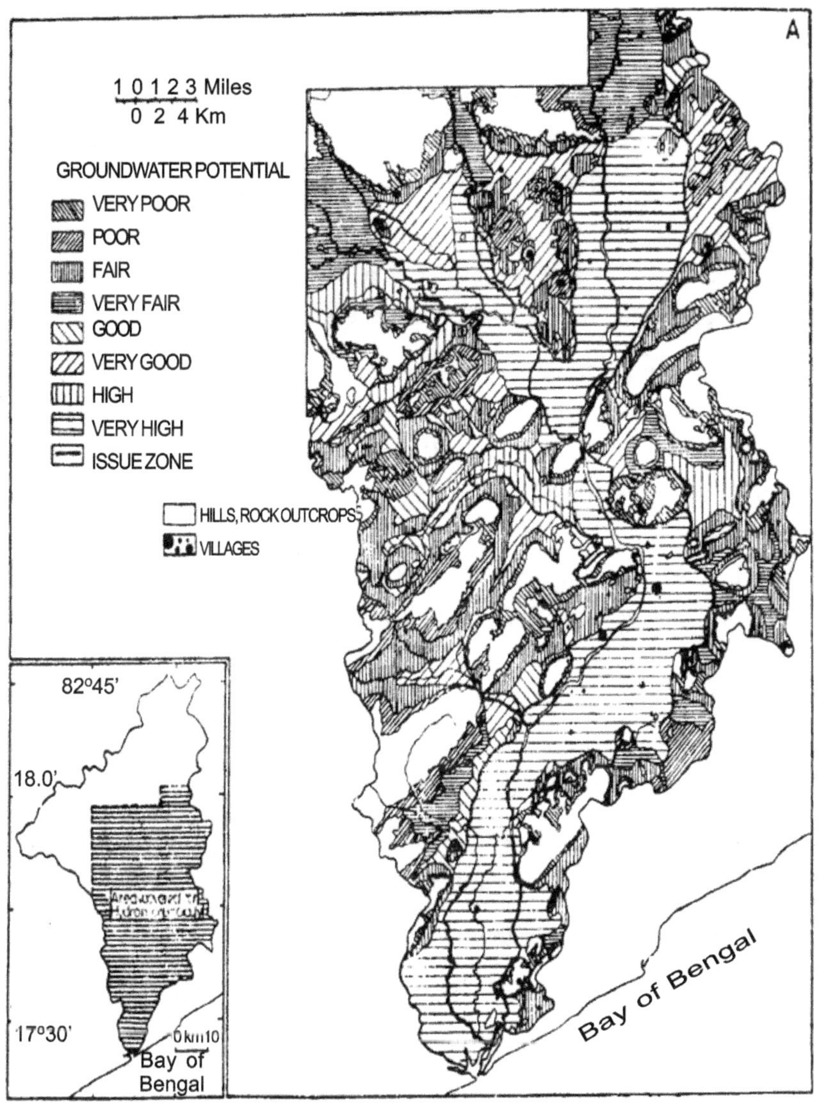

Fig. 3.4 Ground water potentials in part of Sarada river basin.

5. ***Thickness :*** 2 to 4 metres; greater (upto 6 metres) over alluvial fans.

6. ***Other :*** associated with alluvial and coalesced alluvial fans; material is one and the same; a thin occasional layer of sand is found in between colluvium in alluvial fan areas.

7. ***Section*** : 2-4 metres of colluvium on top; white to ochre coloured clayey weathered zone of 0.5 metres thickness; fractured and decomposed rock up to 1 m thickness follows the weathered zone.

8. ***Hydrology :*** fair amounts of percolation possible because of deceleration of run-off water and unassorted nature of material in this unit; clayey weathered zone acts as aquitard; water table is confined to fractured saprolite of bedrock; ground water yield is very poor – 3-8 lpm; ground water flow is towards the sandy plain below; only open dug wells work in this unit; wells carry water only during rainy season.

Sandy Plain

1. ***Location :*** starts from the lower boundary of colluvial plain; ubiquitous unit like colluvial plain

2. ***Slope :*** 3-5 degrees; less in granitic and charnockitic terrain; unidirectional and towards the wash plain below.

3. ***Extension :*** narrow to wide; wider than colluvial plain; 2-3 km in width.

4. ***Material :*** red sand and grit with occasional stone lines and lag deposits/gravels.

5. ***Thickness :*** 4-12 metres.

6. ***Section*** : uniformly similar composition with red sand and grit up to a depth of 4-12 metres; white to ochre coloured clay dominates in 1-1.50m thick weathered zone; the fractured saprolitic bedrock of 1-2 m thickness follows the weathered zone.

7. ***Hydrology :*** very good percolation and well drained; quick and high percolation and evaporation because of good capillarity; clayey weathered zone acts as aquitard; water table confined to 1-2 m thick fractured bedrock occurs at 15 m below the ground level; yield is fair from a minimum of 15 to 35 lpm; open dug wells; ground water flow is towards the wash plain which follows; open dug-cum-borewells suggested; open wells dry up after March.

Pediments : rock-cut and buried rock-cut floors

1. ***Location :*** occur mostly in the central part of the basin between Chidikada and Chodavaram, around Ramayogi Agraharam; and, east of Kondakarla lake in the southern part of the basin.

2. ***Slope :*** 2-3 degrees; unidirectional and towards the wash plain

3. ***Extension :*** circular to sub-circular or fan-like in shape; 2-3 km wide and long.

4. ***Material :*** the rock-cut surfaces are covered under sandy silts with fair amounts of clay in the central part and sandy grit with fair amounts of silts in the southern part.

5. ***Thickness :*** 0-3 m in thickness.

6. ***Other :*** small rock-out crops or pinch outs (so called inselbergs) occur in the centre of these pediments. In the central part they are mostly cut in quartz-felspathic rocks and charnockitic rocks; the rock-floor is a uniformly cut and slopes gently away from the inselbergs; there are weathered depressions here and there in these floors.

7. ***Section :*** The thin overburden of max. 3 m thickness is of uniform composition; weathered zone is poor and clayey and is almost absent; fractured or massive rock follows overburden.

8. ***Hydrology :*** there is scope for good percolation; bedrock being shallow and gently sloping away from the outcrops there can be good sub-surface horizontal flow towards the drainage lines; water table is at a depth of 4 m and the depth of water table bgl increases towards wash plain; ground water is poor yielding upto 5-12 lpm; ground water flow is towards wash plain; only open dug wells are suggested in this unit.

Wash Plain

1. ***Location :*** occurs next to sandy plain or next to pediments; a ubiquitous unit between the river built plain and the sandy plain or between pediments and river built plain.

2. ***Slope :*** 1-2 degrees; unidirectional and towards the drainage lines

3. ***Extension*** : very extensive in occurrence.

4. ***Material*** : clayey silts with fair amounts of sand with occasional thin sandy layers and lenses as intercalations.

5. ***Thickness :*** 4-15 m in thickness

6. ***Section :*** the entire thickness of the deposit above the weathered zone is uniform in composition with two to three dark coloured beds (of max. 0.50 m thickness) of clayey silts; dark colour is due to organic matter and soil formation; the weathered zone is fairly thick – 2 to 3 metres and the fractured and saprolitic bedrock is also fairly deeper – 2 to 3 metres.

7. ***Hydrology :*** percolation good but not better than over sandy plain; the upper deposit and the weathered zone serve as good aquitards; the water table at a depth of 5m but the depth bgl increases up to 15m towards river built plain; yield increases with depth up on reaching saprolitic horizon at a depth of 18m; ground water flow is expected to be unidirectional towards the river built plain which follows; water yield is good to very good with a potential of 20-45 lpm.

River Terraces

1. ***Location :*** occur where the rivers after flowing down the hills break into plains just below the northern hills; they are three in number; the first and topmost at 20m and the second at 15 metres and the third at 3 metres from the river beds.
2. ***Slope :*** 2-3 degrees; unidirectional and it is towards the river flow.
3. ***Extension :*** quite extensive covering a vast area.
4. ***Material :*** white to yellow clayey sands and dark coloured clayey silts alternate.
5. ***Thickness :*** the thickness is about 15-20 metres.
6. ***Section :*** the 15-20 m thick terrace deposits consist of almost alternating layers of white to yellow clayey sands and dark coloured clayey silts with occasional to frequent occurrence of coarse sandy channel deposits; a bed of boulders underlie the terrace deposits near to hills; the clayey weathered zone and the saprolitic zone are 2-3 metres thick.
7. **Other :** the topmost terrace at 20m from the river bed level has almost been eroded way with only a few remnants of it existing; the terrace plain gradually mixes up with the river built plain and the contact is difficult to demarcate.

8. ***Hydrology :*** there is good scope for percolation; water table is at a depth of 16 m which increases upto 20m; water yield increases with depth; water yield is good to very good; yield ranges from 35-75 lpm; ground water movement is in the direction of river flow.

River built Plain : the floodplain

1. ***Location :*** along the major river Sarada and its major tributaries
2. ***Slope :*** it is 1-2 degrees and multidirectional towards the river banks and towards the river flow.
3. ***Extension :*** quite extensitive along Sarada river; less extensive along its major tributary.
4. ***Material :*** white to yellow clayey sands and exclusive coarse sands with grit and gravel.
5. ***Thickness :*** 20-30 metres.
6. ***Section :*** it is a uniform deposit with exclusive sand as thin sheets and thick beds occurring frequently; dark coloured clayey sands as thick beds are not a rare occurrence; the weathered zone (1-3m thick) is clayey and the fractured saprolitic zone is as thick as the weathered zone.
7. ***Hydrology :*** good scope for percolation; water table is at a depth of 5 metres; water yield increases with depth; ground water movement is towards the drainage lines as well as towards the sea; borewells up to a depth of 36-40 metres suggested for very good yield of 45-105 lpm.

Abandoned (buried) Channels

1. ***Location :*** over river built plain and terrace plain.
2. ***Slope :*** sloping towards the present river flow

3. ***Extension :*** narrow and long and quite extensive occurrence in the river built plain.

4. ***Material :*** white to yellow clayey sands at the top just like over river built plain.

5. ***Thickness :*** 2-25 metres.

6. ***Section :*** Fining upward sequence of with coarsest sands and gravels below and fines occupy the upper layers.

7. ***Hydrology :*** water table is at a depth of 5 metres; yield amounts to 80-145 lpm; ground water movement follows the channel course.

Issue Zones

Contacts between ...sandy plain and wash plain (20-50 lpm)

...pediments and wash plain (15-35 lpm)

...wash plain and river built plain (25-65 lpm)

These contacts have clear breaks of slopes and have a clear expression in air-photos and satellite imagery. They have greater yield potential than over different plains/units;

Fractures/Fracture Traces : lineaments

Over these fractures the borewells gave yields upto 60-120 lpm at a depth ranging from 35 to 60 metres.

Caution

The aquifers in the area do not sustain continuous and long periods of withdrawal of ground water.

Recharge

The recharge in colluvial plain, sandy plain is very poor. Quite often the bores go dry in summer season. The wash plain and the terrace plain have somewhat a better recharge condition. The recharge in the river built plain, over abandoned (buried) channels and fracture traces is excellent.

Conclusion

There is a particular hierarchy in the occurrence of various geomorphic units and each unit has a unique surface expression and sediment composition. This hierarchy is consistent in the whole of khondalitic-charnockitic terrain and Eastern Ghats in NE Andhra Pradesh. The fractures cut across different geomorphic units and landforms and may act as conduits for water movement from one unit to another. There is fining of deposits from bottom upwards and towards the drainage lines because of deposition by gravity creep and by ephemeral streams and sheet floods in a semi-arid to arid climate. Since the bedrock is shaped by processes of pedimentation to a uniformly cut and sloping surface towards the drainage lines, the ground water movement follows the slope of bedrock.

4

Geophysics in the Study of Coastal Salinization

I. Radhakrishna

Formerly Scientist, National Geophysical Research Institute, Hyderabad-7, A.P., India

Key words : Coastal groundwaters, Salinization, S.P. and Resistivity logs, Hydrogeochemistry

Introduction

The chapter explains the ways and means of using the geophysical (principally self-potential and electrical resistivity) methodologies to monitor and mitigate salinization of coastal groundwaters and soils, based on Indian case histories.

Coastal ground water systems constitute a very important resource for drinking, irrigation, agriculture and industrial sectors of the country and play a vital role in the growth and development of the rural base. These coastal systems are interfaced with oceanic basins on one side and upland regions with major river valley projects on the other end and become targets for subsurface hydrodynamic and hydrochemical changes on either side causing saline water ingression inland into the fresh water systems. Physico-chemical processes accompanying such ingression, aquifer reservoir characteristics and paleohydrogeological processes in different sub-systems of the deltas are the major elements required to be studied and quantified. These would help to develop appropriate orptimization schemes, models with predictive capabilities on physico-chemical changes, their space, time variations and protective measurers for sustained development and management of fresh ground water resources.

Over the last two decades or so the fresh ground water resources in coastal deltas have been subjected to large scale stresses in the form of ground water withdrawal, resulting in reduction of water levels and fresh water storage potentials causing saline water ingression processes inland into fresh water systems. This situation is being further threatened by global warming processes and possible sea level rise events in future which may further deteriorate the situation. It is therefore vitally necessary to obtain the first hand information on the existing fresh ground water systems, their areal spread and boundary conditions and their inter-relationship with adjoining saline water bodies. These are site specific problems and are required to be studied basin wise all along the coast in different deltas and interdelta regions. The most important problem in the Indian costal delta system is identification of interface/diffusion boundaries. Fresh/saline water interface boundary is a complex and dynamic phenomenon and could occur in a variety of combinations and shapes such as : (a) Fresh water aquifers either overlying or underlying saline water aquifers (b) Fresh water aquifers sandwiched between two saline water aquifers or vice versa (c) Alternate system of fresh and saline water aquifers and (d) Fresh water aquifers wedging towards the coast or saline water a aquifers wedging towards the continent. All these complex combinations of saline/fresh water aquifers make the well field development strategies rather difficult and demands very accurate data structure and analyses in relation to: 1. thicknesses of fresh/saline aquifers and their interface boundary. 2. nature of interface or diffusion boundary 3. estimations of wedging of saline/fresh water aquifer and their geometrical variations 4. estimations of depth levels for concentrate plugging in the gravel packs to avoid vertical saline water inflows and mixing and 5. designing appropriate monitoring observational network particularly near the wedge shaped aquiclude boundaries where fresh and saline water systems come into juxtapositions and 6. assessment of possible vertical leakages through aquitards from fresh aquifers to saline aquifers and vice versa. All the above described problems call for precise estimations of subsurface hydrogeological boundaries and interestitial water quality characteristics as precisely as possible, so that the well field development could be planned accurately. Though several techniques are available like surface geophysics, remote sensing, hydrogeological and hydrogeochemical and packer tests, none of these methods help us to accurately quantify the depth levels of interface/diffusion boundaries, interstitial water qualities and wedging of aquifers and aquicludes. A reasonably powerful tool for estimation of interface/diffusion boundaries and water quality characteristics in multi aquifer systems of coastal deltas is borehole geophysical techniques. (Jones and Buford, 1957).

Methodology Based on Subsurface Geophysical Profiles

Self potential and electrical resistivity logs have been widely used in a qualitative way to define the bed boundaries, their thicknesses, lithological correlations, aquifers reservoir boundaries and also to define saline fresh water boundaries which would serve as an important database for well field development and estimation of mechanism of saline water ingression processes etc. Very limited efforts so far have been made to quantify the water quality parameters in terms of different ionic components in multi aquifer coastal systems. Bore hole geophysical logs involving both electrical resistivity and self potential logs play a vital role in this regard. Estimation of true resistivity of formational waters in both logging procedures is very crucial. Jones and Buford (1957) attempted quantitative interpretation of resistivity data and obtained reasonably good results. Resistivity of formation water (R_w) has a definite relation to dissolved solids and one can estimate R_w values by taking into consideration the formation resistivity factor. Kwader (1986) estimated R_w values for Tertiary carbonate and granular formations by cross plotting saturated formation resistivity (R_0) against bulk porosity. These data, when plotted on Hingle porosity cross plots yielded graphical solutions for estimations of R_w values. Alger and Harrison (1989), using true formation resistivity (R_t), flushed zone resistivity (R_{xo}), mud filtrate resistivity (R_{mf}) and spontaneous potentials presented solutions to quantifying formation water quality, permeability (hydraulic conductivity) and effective porosity (storativity). Spontaneous potential logs can also provide comparable evaluations of hydrogeochemical parameters for different zones, however, the number of data and case studies available is rather limited. Alger (1966) demonstrated the usefulness of Self Potential (SP) logs for Gulf Coast aquifers and obtained good results. Vonhoff (1966) found a workable empirical relationship between spontaneous potential deflection on the electric log and water quality in glacial aquifers. Dyck et al. (1972) also stressed the importance of SP logs based on their studies on Kipling sand aquifers. Hearst and Nelson (1985) made critical analyses and reviewed major approaches towards solving many groundwater problems including interpretation of SP curves. In view of the non-availability of good case studies on spontaneous potential log interpretation for water quality problems an attempt is made in this paper to evaluate the water qualities in a coastal groundwater system in the state of Orissa in India.

The True resistivity of a formation in general through an electrical resistivity log could be obtained by using the following equation (Archie 1942).

$$R_t = F * R_w \qquad .....(4.1)$$

Where R_t = true formation resistivity;

R_w = saturated water resistivity of an aquifer;

F = formation factor which is inversely proportional to the porosity raised to an exponential power that represents void distribution.

The true resistivity of the formation and water saturated zone could be obtained if the data related to electrode spacing, mud resistivity, diameter of the borehole, the apparent resistivity of the bed and formation factors are available. Estimations of true water resistivities through the above processes and parametric estimations are rather complicated and could lead to several errors during the interpretational analysis. It is therefore the endeavour of this paper to present and demonstrate a more rational and reliable method using self potential logs and present the interface structure for Mahanadi - Brahmini-Baitharni delta basins in Orissa along the Indian coast. The self potential (Sp) log is a response of a log profile in general affected by several factors like borehole diameter, bed thicknesses, shalyness of formation, quality of the formation water, mud filtration invasion and temperature. The electrochemical potential generated in such logs is given by the formula :

$$Sp = -k \log R_{mf}/R_w \qquad .....(4.2)$$

where Sp = log deflection (mV) against a zone ;

k = a factor generally taken to be equal to 70, but deviates marginally with temperature; 60 + 0.133T

where T = borehole temperature (^oF);

R_{mf} = Resistivity of mud filtrate (Ωm^{-1});

R_w = Resistivity of formation

The above equation is essentially based on the following assumptions : (i) both the borehole fluid and formation waters are sodium chloride solutions; (ii) the shale formations are ideal ion selective membranes and the sand formations have no ion sieving properties; (iii) the borehole fluid has a much greater resistivity than the combined resistivity of the sand and shale (Keys and McCary, 1971). The above relationship is used effectively in the oil industry to estimate R_w (resistivity of water) through Sp logs since NaCl is the dominant salt in solution and the interrelationships between concentration, activity and resistivity are well established. This can't be applied directly in the water industry because the medium generally is not dominated by NaCl waters. To avoid this constraint the following method is adopted to evaluate and quantify the water quality parameters for unknown zones at depths through Sp-curves.

1. A number of sites are selected for which water analyses and electrical log data are available. A linear relationship between R_w and total dissolved solids (TDS) for the ground water system under consideration is obtained. (Fig. 4.1.)

2. The R_w - R_{we} interrelationship for NaCl waters and $NaHCO_3$ waters are plotted on double log scale and the factor of deviation is obtained. (Fig. 4.2.)

3. Based on analytical results selected for a set of samples, scatter trend diagrams between TDS land different anions and cations are prepared. These graphical trends serve as standard diagrams for the area under consideration. (Figs. 4.3 & 4.4.)

4. R_w values for unknown zones are calculated and TDS and other ions are interpreted with the help of graphical trends. The resolution for interpretation of TDS values appears best for aquifers with R_w values between 5 ohm/m to 20 ohm/m and the accuracy of interpreted data gets reduced beyond these boundaries. In such cases the data interpretation is made using the formula given below.

$$R_w = 10{,}000 \,/\, \text{Sp.conductance} \qquad .....(4.3)$$

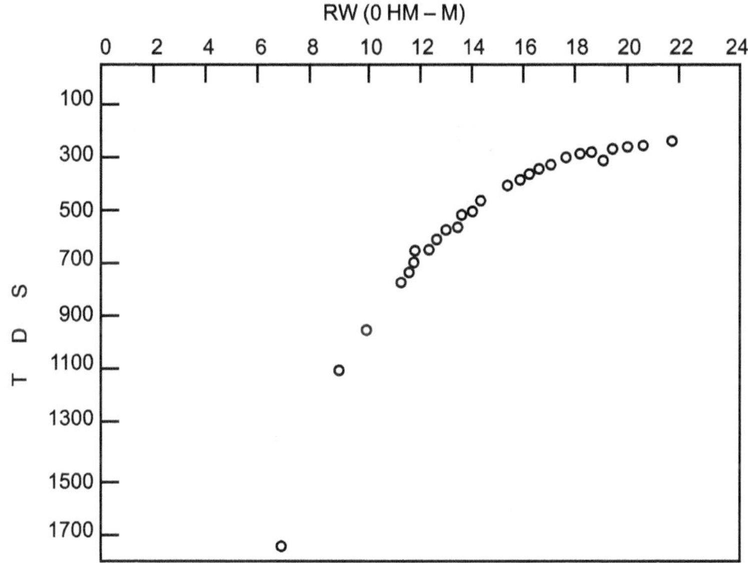

Fig. 4.1 Relation between TDS and RW.

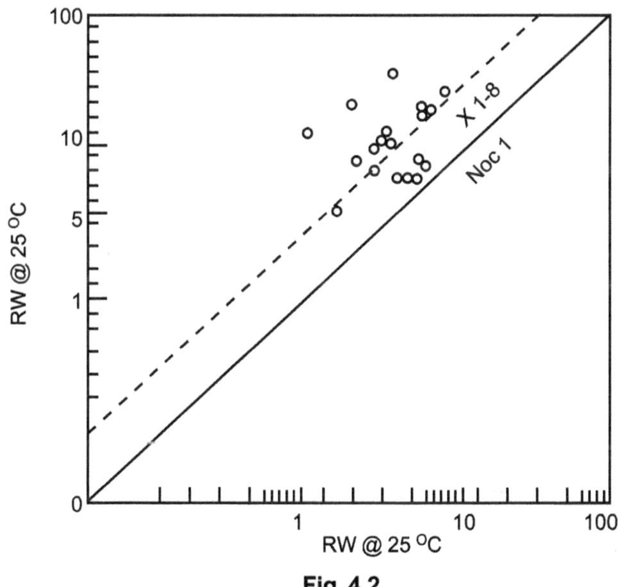

Fig. 4.2

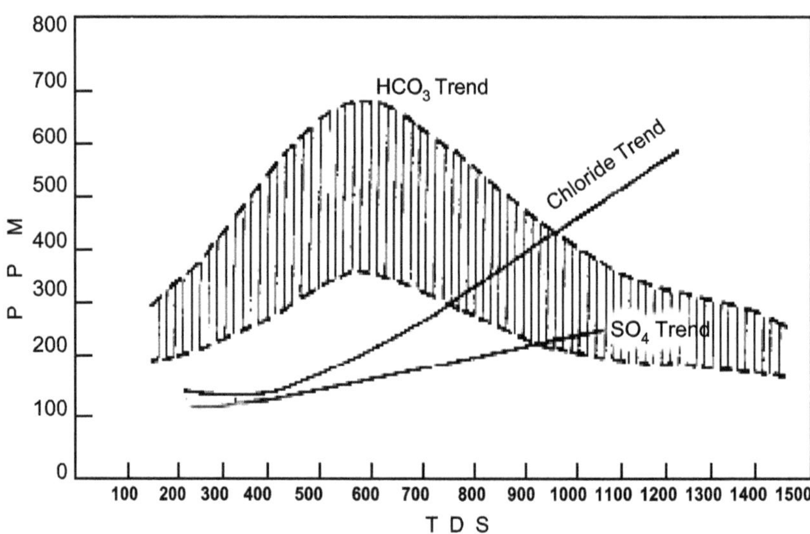

Fig. 4.3 Chemical activity of groundwaters (Orissa).

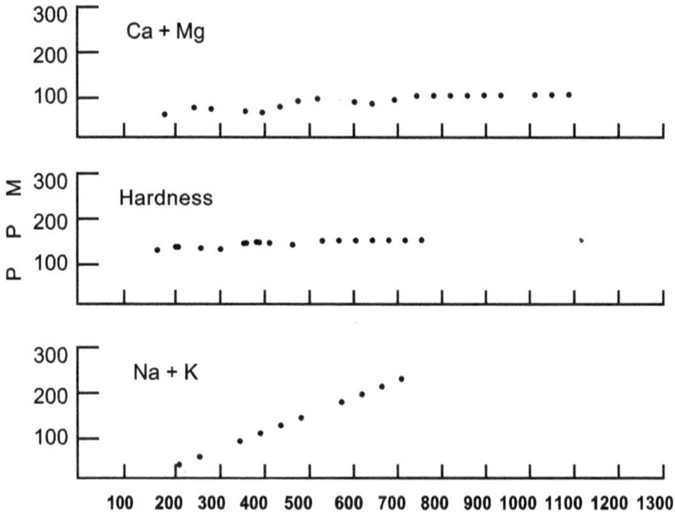

Fig. 4.4 Chemical activity of groundwaters (Orissa).

Derivation of Hypothetical and Actual Chemical Analysis

1. Based on the vertical and lateral changes of water qualities in a ground water system, standard graphical interrelationship between R_w (true resistivity of water) and total dissolved solids, has been built for coastal Mahanadi-Baitarni-Brahmini complex along the Indian coast (Fig. 4.l). This is possible when several water analyses of ground water samples from selected wells and their R_w values, are available.

2. The R_w-R_{we} relationship for NaCl waters is given by a diagonal line in Fig. 4.2. The R_{we} values using eqn. 4.1 and the R_w values obtained using chemical analyses in the coastal groundwater system are plotted against each other and a straight line is drawn by least square fitting, which is displaced upwards such that for every value of R_{we}, the R_w value is 1.80 times greater than for NaCl solutions. The predominant cations and anions in solution are sodium and bicarbonate. The activity of the solution is primarily dependent on cation concentration and the R_w values of NaCl and $NaHCO_3$ values are different. The derived R_w values for the groundwaters of different zones at each of the well sites are given in Table 4.1.

3. The trends of the scatter plots between TDS and various cations and anions are shown in Figs. 4.3 and 4.4. Almost all the ions are represented by linear trends with different slopes expect bicarbonate values which are represented by a field with certain boundary values.

The above methodology developed had helped to build the quantified water quality parameters for both shallow and deeper aquifer zones in a complex coastal system like Mahanadi-Baitarni-Brahmini deltas in India. Based on these results an interface configuration map for the entire region has been prepared and the saline water intrusion processes/mechanisms have been explained taking the flow dynamics into consideration. This methodology perhaps could be adopted in similar systems elsewhere in the world.

Table 4.1 Results obtained from self potential log analyses

Site and Location	Nature of aquifers	Zone depth(m)	Sp	R_{mf}	R_{we}	R_{mf}/R_{we}	R_w
South Mahanadi Delta Sp log data for shallow aquifer systems							
Arilo (long. 86°13'38" Lat. 20°12'15"	Fresh	0-60 60-96 96-107	+ 22.5 + 12.5 – 15.0	6.50 6.50 6.50	13.63 9.81 3.97	0.48 0.66 1.64	24.525 17.650 7.143
	Brackish	144-167 188-210 220-245	– 15 – 25 – 40	6.50 6.50 6.50	4 2.890 1.778	1.625 2.249 3.656	3.8 2.8 1.8
Machagaon (Long. 86°16'0" Lat.20°4'0"	Fresh	0-50 50-100	+ 22.5 – 25.0	5.00 5.00	10.48 2.20	0.48 2.27	18.865 3.954
	Brackish	108-146 172-193 196-222	– 40 – 50 – 50	5.00 5.00 5.00	1.368 0.988 0.988	3.655 5.06 5.061	1.4 10.98 0.98
Patalia (Long. 86°2'35" Lat. 19°57'24"	Fresh	20-40 40-80 80-110	+ 10.0 0.0 – 7.5	6.00 6.00 6.00	8.34 6.00 4.69	0.72 1.00 1.28	15.006 8.430 8.430
	Brackish	122-138 146-179	– 45 – 50	6.00 6.00	1.394 1.186	4.304 5.059	2.509 1.186
North Maha Sp log data							
Jajang (Long. 86°25'52" Lat. 20°31'24")	Brackish Fresh	85-105 105-250	– 47.5 – 15.0	5.90 5.59	1.24 3.60	4.76 1.64	2.226 6.483

Table 4.1 Contd......

Masakani	Brackish	22-55	– 20	3.15	1.646	1.914	2.692
(Long.		58-75	– 15	3.15	1.937	1.626	3.487
86°35'50"							
Lat.		105-150	+ 20	3.15	6.029	0.522	0.852
20°24'30")		274-286	– 20	3.15	1.646	1.914	2.692
	Fresh	290-310	– 13	3.15	2.067	1.524	3.720
Adjhori	Brackish	36-44	– 30	6.00	2.265	2.649	4.077
(lang.86°45'		52-62	– 50	6.00	1.186	5.059	2.135
Lat.20°34')							
	Fresh	65-95	+ 10.0	6.00	8.34	0.72	15.006
		96-140	+ 14.6	6.00	9.63	0.62	17.330
		140-155	+ 29.0	6.00	15.37	0.39	27.760
		165-220	– 7.5	6.00	4.69	1.28	8.430
Brahmini-Baitarni Delta							
Kherang	Brackish	120-140	– 50.0	9.50	1.83	5.19	3.301
(Long.86°45'							
Lat.	Fresh	150-165	– 25.0	9.50	4.17	2.28	7.513
20°59'30")		180-190	– 20.0	9.50	4.92	1.93	8.356
		190-210	– 20.0	9.50	4.96	1.91	8.900
		220-240	– 30.0	9.50	3.54	2.68	6.374
Mandari	Fresh	18-30	– 30.0	13.00	4.9	2.642	8.82
(Long.'		55-71	– 35.0	13.00	4.17	3.112	7.51
86°47							
Lat.20°15')		106-124	– 15.0	13.00	8	1.625	14.4
		154-203	– 15.0	13.00	8	1.625	14.4
Dakshinabad	Fresh	43-53	– 30.0	11.25	4.258	2.642	7.664
(Long.		89-107	– 50.0	11.25	2.224	5.058	4.003
Lat.20°54')		165-198	– 50.0	11.25	2.224	5.058	4.003
		199-217	– 45.0	11.25	2.613	4.305	7.749
Musanga	Brackish	38-49	– 50.0	3.50	0.692	5.058	1.246
(Long.'		53-60	– 50.0	3.50	0.692	5.058	1.246
86°38							
Lat.21°38')	Fresh	110-148	– 50.0	3.50	2.529	1.384	4.552
		155-214	– 50.0	3.50	1.829	1.914	3.445
		278-287	– 50.0	3.50	1.829	1.914	3.445

Table 4.2 Hydrogeochemical parameters (laboratory and interpreted value)

	Values based on laboratory Analysis of water samples (ppm)								Values interpreted from sp logs (ppm)							
Well site and zone Depth (m)	**Sp. Cond. (µmhos cm^{-1})**	**TDS**	**Cl**	**SO_4**	**Hco_3**	**Ca+ M g**	**Na+k**	**TH[1]**	**Sp. cond (µmhos cm^{-1})**	**TDS**	**Cl**	**SO_4**	**Hco_3**	**Ca+ M g**	**Na+ k**	**TH[1]**
Hydrogeochemical parameter for shallower/deeper ground water System at selected wells/zones South Mahanadi Delta																
Arilo (69-90) F	725	471	44	-	232	45	-	135	560	365	38	20	150-372	50	70	150
(188-210) B	-	-	N.D.A	-	-	-	-	-	2640	1650	540	260	100-160	90	500	180
Patalia F (80-110)	1041	677	142	-	329	16	-	146	1200	780	200	95	60-387	60	220	160
(122-138) B	-	-	N.D.A	-	-	-	-	-	4160	2600	960	300	20-70	140	750	200
Machagoam (0-50) F	747	485	47	-	232	34	-	105	525	340	30	165	140-35	48	83	173

Table 4.2 Contd......

Values based on laboratory Analysis of water samples (ppm)									Values interpreted from sp logs (ppm)							
Well site and zone Depth (m)	Sp. Cond. (μmhos cm^{-1})	TDS	Cl	SO_4	Hco_3	Ca+ Mg	Na+k	TH[1]	Sp. cond (μmhos cm^{-1})	TDS	Cl	SO_4	Hco_3	Ca+ Mg	Na+ k	TH[1]
(108-146)	-	-	N.D.A	-	-	-	-	-	3360	2100	730	340	40-100	110	800	198
B																
Hydrogeochemical components for shallower/deeper groundwater systems at selected wells/zones																
North Mahanadi Delta																
Jaiang (115-200)	1648	707	199	-	115	28	-	135	1570	1020	323	140	100-253	60	300	160
F																
(85-105)	-	-	N.D.A	-	-	-	-	-	4480	2800	1040	420	20-50	170	750	215
B																
Masakani (274-286)	-	-	N.D.A	-	-	-	-	-	3840	2400	860	340	30-80	98	740	230
B																
(290-310)	-	-	N.D.A	-	-	-	-	-	2784	1740	570	260	50-125	80	560	180
F																
Adhori B(58-75)	-	-	N.D.A	-	-	-	-	-	2912	1820	600	280	55-100	100	700	175
(165-200)																

Table 4.2 Contd......

Values based on laboratory Analysis of water samples (ppm)									Values interpreted from sp logs (ppm)							
Well site and zone Depth (m)	Sp. Cond. (μmhos cm^{-1})	TDS	Cl	SO_4	Hco_3	Ca+ Mg	Na+k	TH[1]	Sp. cond (μmhos cm^{-1})	TDS	Cl	SO_4	Hco_3	Ca+ Mg	Na+ k	TH[1]
F	983	639	180	-	165	31	-	205	1200	780	200	95	60-387	69	220	160
Brahmini-																
Baitarni																
Delta																
Kherang	-	-	N.D.A	-	-	-	-	-	3136	1960	670	280	40-100	110	700	180
B(120-140)																
(190-210)																
F	830	540	115	50	161	17	135	141	1015	660	150	77	195-450	52	190	160
Mandari	-	-	N.D.A	-	-	-	-	-	704	440	70	30	160-395	40	110	150
(106-198)																
F																
Dakshinabad	-	-	N.D.A	-	-	-	-	-	2892	1620	520	200	100-160	90	470	170
(165-198)																
F																
Musanga	-	-	N.D.A	-	-	-	-	-	3008	1880	640	310	58-100	110	580	192
F(236-272)																

Th[1] = total hardness; B = Brackish Water; F = Fresh Water; N.D.A = No Data Available

Saline Fresh Water Interface Structures

The interface structure, nature of interface and boundary conditions obviously become very important parameters in most coastal regions of the world. This would form the basis for well field development and also to understand the mechanism of saline water intrusion processes. In order to prepare an interface structure map of Mahanadi-Baitarni-Brahmini region, about 12 representative well sites have been selected (Fig. 4.5) where the interface boundaries are clearly established. The nature and characterstics of interface at each well site are shown in Table 4.3.

Based on the data obtained from these well sites and available log parameters an interface structure map for the entire region has been prepared and presented in fig.6 while accounting for the parameters like the depths and nature of the interfaces. The interface could be a hydraulic diffusion boundary or a geological aquiclude/aquitard boundary separating these two fluid systems. Each of these boundaries have a specific role to play towards designing the well field structures. The following table gives results of such analysis.

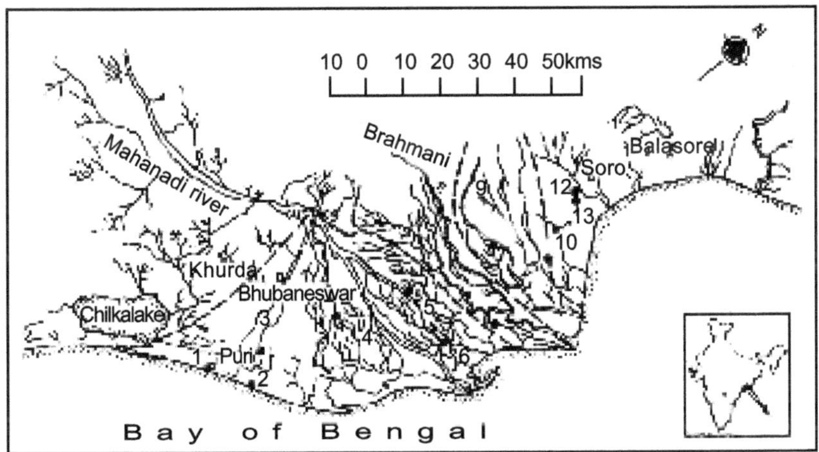

Well Site Locations

1. Puri 2. Konark 3. Patalia 4. Arilo 5. Jajang 6. Masakani 7. Dubakona 8. Madanpur 9. Dakshinabad 10. Musang 11. Kherang 12. Mandari

Fig. 4.5

Table 4.3 Saline fresh water interface characteristics at different well aites

Sl. No.	Name of the well site	Inter-face depth (m)	Nature of interface	Remarks
1.	Puri	20	Diffusion Zone	Fresh water over brackish water
2.	Konark	30	Diffusion Zone	Fresh water over brackish water
3.	Patalia	100	Diffusion Zone	Fresh water over brackish water
4.	Arilo	100	Diffusion Zone	Fresh water over brackish water
5.	Jajang	120	Aquiclude Zone	Brackish water over fresh water
6.	Masakani	290	Aquiclude Zone	Brackish water over fresh water
7.	Dubakona	50	Aquiclude Zone	Brackish water over fresh water
8.	Madanpur	60	Aquiclude Zone	Brackish water over fresh water
9.	Dakshinabad	Not appli-cable	Not applicable	No salinity
10.	Musanga	60	Aquiclude Zone	Brackish water over fresh water
11.	Kherang	60	Aquiclude Zone	Brackish water over fresh water
12.	Mandari	Not appli-cable	Not applicable	No salinity

The interface structure map presented here (Fig. 4.6) demonstrates the occurrence of fluid systems in three main categories. 1. Fresh water aquifers underlain by a brackish water aquifers with or without a diffusion boundary; 2. Fresh water aquifers overlain by brackish water aquifers with aquiclude/ aquitard boundary; 3. Alternate system of fresh - saline water aquifers.

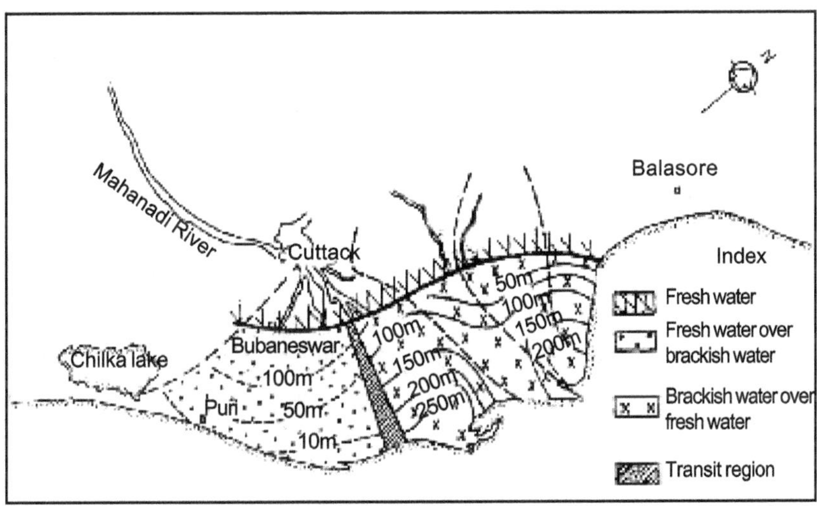

Fig. 4.6 Interface depth structure map for Mahanadi - Brahmini - Baitarani basins : Orissa

South Mahanadi Delta

The region south of river Khatjori (a tributary to Mahanadi) and extending almost upto the Chilka lake of the delta alluvium falls under this region. Fresh water aquifers occupy the top slots of the aquifer regime with depths varying from 10m - 100m. The thicknesses of the fresh water regime increases upstream and reduces towards downstream with interface zone normally represented by diffusion boundary. Any ground water development therefore should take into account the upcoming phenomenon while designing the well field structures.

North Mahanadi Delta

This region falls between Birupa and Mahanadi interregion. Unlike south Mahanadi delta the top slots of the aquifer regime are occupied by brackish water aquifers on the top underlain by fresh water aquifer systems with high pressure heads. The interfaces are normally represented by thick aquicludes, which gradually pinch towards central and upstream areas of the delta with depths varying from 60 m - 280m. The interface boundary steeply deepens towards Mahanadi outfall area and shallows up towards upstream side and Birupa- Brahmini outfall areas. It becomes therefore necessary in such systems to adopt cement plugging in the gravel pack area at different depths of interfaces to avoid vertical migration while pumping. It is also necessary in such situations to estimate the lateral extensions of aquiclude interface boundaries upstream where saline-fresh water systems come to juxtapositions and optimize/ control the pumping discharges in such a way that the piezometric cone should not spread beyond this critical zonal limit failing which the saline water gradually spreads into bottom areas of fresh water and could convert the fresh water regime into brackish water regime.

Brahmini-Baitarni Delta

This region by and large is represented by the same trends as that of north Mahanadi delta with saline water on the top underlain by fresh water aquifers separated by aquicludes. The interface depths in this region varies from 60m - 150m with relatively shallow interface depths in Brahmini delta and deeper interfaces in Baitarni delta. The same operational procedures are required to be adopted in this region also while designing the well field developmental structures.

Conclusions

1. The self potential logs if obtained accurately help us to estimate not only the bed boundaries, thicknesses etc., but also to quantify the ionic components in a multi aquifer system which will have a great bearing on the well field development and source and mechanism of saline water intrusion processes.
2. The interface structural map prepared for Mahanadi-Brahmini-Baitarni region demonstrates that the interface/ diffusion boundaries need not necessarily follow the existing hydrodynamic principles but appear to be influenced by paleo-sedimentation hydrological processes and basin structure. The map also suggests that the outfall regions of Mahanadi and Baitarni are the principle source areas of saline/brackish water intrusion processes.
3. Design of well field developments in a complex situation like this as described above requires quantified estimates in relation to "upconing processes" in the case of diffusion boundaries and "lateral spread of peizometric cone" in the case of aquiclude boundaries to check the brackish water migration from one aquifer stage to another.

References

Archie.G.E., Jan., 1942, "The electrical resistivity log as an aid in determining some reservoir characteristics", *Pet. Tech.*, 5 (1)

Alger, R.P., 1966. *Interpretation of electric logs in Fresh Water wells in unconsolidated formation*: Soc. Prof. Well Log. Analysts, 7th Annu. Logging Symposium, Tulsa, OK, May 1966, Trans., R.CC1-CC25.

Alger, R.P., and Harrison, C.W., 1989. Improved fresh water assessment in sand aquifers utilizing geophysical well logs. *The Log Analyst*, Jan.-Feb., 1989, pp. 31-44.

Dyck, J.H., Keys, W.S. and Meneley, W.A., 1972. Application of geophysical logging to ground-water studies in southeastern Saskatchewan. *Can. J. Earth Sci.,* 9(1): 78-94.

Gondouin, M., Tixier, M.P. and Simard, G.L., 1957. An experimental study on the influence of the chemical composition of electrolytes on the SP curve. *J. Pet. Tech.,* 9(2), SPE reprint series, 1971, No. 1, pp. 65-79.

Hearst, J.R. and Nelson, P.H., 1985. *Well Logging for Physical Properties*. McGraw-Hill, NY.

Jones, P.H. and Buford, J.B., 1957. Electric logging applied to Groundwater Exploration. *Geophysics*, 16(1): 115-139.

Keys, W.S. and McCary, L.M., 1971. *Application of borehole geophysics to water resources investigation.* Book 2, Techniques of Water Resources Investigations of the United States Geological Survey.

Kwader, T., 1986. The use of geophysical logs for determining formation water quality. *Ground-water*, 24(1): 11-15.

Vonhoff, J.A., 1966. Water quality determinations from spontaneous — potential electric log curves. *J. Hydrol.*, 4: 341-47.

5

Multi-Thematic Remote Sensing in Water Resources Management

D.P. Rao

Former Director, National Remote Sensing Agency,
(Department of Space, Government of India), Hyderabad 500 037, India
Email: dprasad@hotmail.com

Key words : Remote sensing applications, Management, Surface water resources, Groundwater Prospect maps, Integration with geology and geomorphology.

Introduction

The ever-growing world population with expanding activities in agriculture and industry resulted in increased demands on the limited water resources. There has been an annual increase in the water withdrawal of between 4 and 8 percent, with highest rate of growth occurring in developing countries (WRI, 1992). Furthermore, pollution by chemicals, radioactive materials, and suspended solids pose a further threat to security and quality of water supply. Realizing the importance of water resources management, the UN convened a Water Resources Conference in 1977. The recommendations of the conference have been known as Mar del Plata Action Plan (MPAP). Periodic reviews made in implementing the action plan have stressed the need for a more integrated and inter-sectoral approach to water resources management than has hitherto been the case (WMO/UNESCO, 1991; FAO, 1990;WHO/UNEP, 1991). Inadequate supply of water for irrigation is one of the major issues of water resources management in arid and semi arid regions especially in the Asia-Pacific region. The incidence of water logging and subsequent development of soil salinity and alkalinity is an equally serious problem in irrigated lands.

The global and regional water resources scenarios highlight the need to find improved ways to meet the problems of food security, water availability and water use efficiency. The quality of the land and water must be sustained in the face of mounting pressure of water logging, soil salinization, groundwater depletion and water pollution. The coupled impact of such increasing population and water consumption on the global ecosystem may also lead to climate change which may contribute to and constrain future development (Kasturirangan, 1996). The ongoing conservation practices alone will not be enough to meet the type of challenges we would face in future. Identification and adoption of appropriate technologies to overcome the pressure is, therefore, a pre-requisite. This necessitates broad thinking and drawing up a research agenda that would address these concerns and would contribute towards agriculture productivity, sustainable use of groundwater and reduce the danger of malnutrition and health. Integrated river basin management, which focuses on inter-sectoral management of water rather than sectoral management would be the key strategy.

Global Scenario

The annual renewable fresh water supply in the world according to a UN study is about 47000 cubic km. Globally 3240 cubic km of fresh water is withdrawn and used annually. Agriculture uses about 69 percent of World's fresh water with a third of global food production coming from irrigated agriculture while industries use about 23 percent and 8 percent for domestic use (Shiklomanov, 1998). The preservation of natural environment also critically depends on water. Against these requirements, fresh water constitutes less than 2 percent of the total water availability, with lakes and rivers accounting for less than 0.014 percent of all water. Global population is expected to reach 8 billion by 2020, with most of the increase being in the developing countries. This will lead to decline in per capita availability of arable land and clean water. About 12% of the world's total land surface is used presently for cultivation of crop species. According to FAO, an additional 200 million ha. new crop land would be required over the next 30 years to feed the increasing population, given the available food production trends.

Water is critically scarce in 22 countries where the renewable water resource is less than 2000 cubic metres per person. Falling water tables from over pumping of ground water now are ubiquitous in many countries. The present ground water withdrawal according to a UN study is to the tune of 3750 km^3/yr of which only 2270 km^3/yr are actually consumed globally. This is growing by about 10-12% every decade. It is anticipated that by 2025, the water withdrawal would reach about 5100 km^3/yr with about 1.4 times increase over the present. This would result in most of the world's population facing catastrophically low water supplies (Shiklomanov, 1998).

The Indian Scenario

The annual precipitation of India is estimated at 4000 cubic km, with 75 percent of it received during the southwest monsoon season. The ultimate surface water resource is estimated at 1869 cubic km. With the increase in population there has been a decline in per capita availability of water.

An estimated 215 M ha.m of water from rainfall infiltrates into the ground every year. The Central Ground Water Board has estimated the net utilizable ground water resource at around 42.7 M. ha.m. While the total availability of water presently shows a comforting picture, its spatial and temporal distribution, however, is incompatible with the consumption patterns. There is thus an urgent need for planning and operating irrigation schemes, water supply to towns and industries, coping with natural disasters such as drought and flood, maintaining acceptable water quality for human and other uses, designing and operating conveyance system, etc. based on reliable and objective assessment of quantity and quality of water resources (Mohile et at., 1996).

Role of Remote Sensing

Water resources management is essentially a linkage between availability from various sources and sectoral demand in which both quality and quantity need to be considered, and where both conservation and control measures need to be addressed. Conjunctive use of surface and ground water is an integral part of water resources management. Both remote sensing and ground measurements have to be integrated to provide an overview of large area as well as details of local availability. Conventional hydrologic measurements on ground suffer from the limitations of reliability, time effectiveness and adequacy. These measurements are also discrete in space, necessitating areal averaging methods. Repetitive ground measurements many times are scarce due to the constraints of manpower and funds. Measurements over inaccessible areas and inhospitable terrain are also a limitation. It is in this context that remote sensing technique can complement and supplement ground measurements, to enable collection of data needed for sound water resources management.

Remote sensing has been used as an important tool in the inventory, survey, monitoring, planning and management of natural resources including water resources. The steady flow of data from the constellation of India

Remote Sensing Satellites (IRS-1A,-1B,-P2,-1C,-P3,-1D and P4) and other foreign missions viz.SPOT-3&4 and LANDSAT –5&7, ADEOS, ERS–1&2, JERS and RADARSAT have facilitated operational use of this technology in the management of natural resources including water resources by the Department of Space, Government of India. An attempt has been made in the following sections to provide a glimpse of the applications of remote sensing in some important areas of water resources management.

Rainfall, Snow and Glacier Studies

Annually, India receives about 4000 cubic km. of water through precipitation : 95% from rain and 5% from snow and glacier melt. While rain is a good source of fresh water, it is a very poor distributor. Unlike in some parts of the world, the distribution of rain varies spatially and temporally in India. The analysis of more than 120 years of monsoon rainfall data for the country from 1871 to 1993 reveals that four major climatic epochs namely, (i) 21 years (1878-1898) of increasing tendency above normal rainfall of 850mm (ii) 30 years (1898-1932) of decreasing tendency (below normal rainfall) (iii) 34 years (1933-1964) of increasing tendency (above normal) and (iv) 23 years (1965 - 1987) of decreasing tendency.

Rainfall is one of the most important processes in the hydrological cycle and is also one of the most difficult to monitor. Since late 1960s, many researchers have attempted to derive techniques for the estimation of rainfall from the visible and infrared imagery provided by meteorological satellites. Manual, interactive and automatic methods have been developed for the estimation of rainfall at a number of temporal and spatial scales, and these have been applied in many different areas and situations with varying degree of success.

Snow acts as a water reservoir. Hence, assessment of its possible effect on snowmelt runoff is essential for planning and management of various multipurpose projects on snow-fed rivers. Unlike in many other countries, where snowfields stretch along flat terrain and mountains that are easily approachable, the snow bound land in India is mostly in difficult and inaccessible mountainous terrain. Remote sensing, therefore, continues to be the only practical way of obtaining information on snow cover. The economical hydroelectric potential of Indian subcontinent has been tentatively assessed at 85550 MW at 60 per cent load factor (Central Electricity Authority, Government of India). Seventy five percent of the potential lies in the rivers of Himalaya, which are fed by snow, offer large hydropower potential on account of topography and perennial nature of flow. Snowmelt runoff estimation is thus, a critical item for planning and management.

Seasonal and short- term (weekly) forecasts of snowmelt runoff are being provided for some basins in western Himalaya by the National Remote Sensing Agency since 1970s with less than 5 per cent error for use by the State River Water Management Board and the State Electricity Board. Inventory of glaciers in the Himalaya have also been conducted at different scales using satellite data, and the information on their numbers and areal extent has been generated as a bench mark study for use in future monitoring.

Irrigation Water Management

In a number of developed countries, the extent of irrigated land has now stabilized or even diminished (Shiklomanov, 1998). At present, about 15% of all cultivated land is being irrigated. In India, the ultimate irrigation potential has been estimated at 113 M. ha. While enormous irrigation potential has been created at huge cost, the gap between created potential and utilization is significantly large (around 9 M.ha). The technology should, thus, shift from irrigation development to more efficient irrigation management. Prolonged irrigation without providing for adequate drainage has led to water logging and subsequent soil salinity and/ or alkalinity in some areas.

Space borne multi-spectral measurements at regular intervals have helped evaluating the performance in many irrigation projects across the country. The anticipated increase in irrigation area, equitable distribution and crop productivity under sponsored development programmes have been studied in some of the major irrigation command projects in. The temporal and spatial analysis of satellite data helped in mapping problem pockets of poor performance.

Diagnostic analysis supported by farmer surveys has enabled identification of causative factors for taking up corrective measures. The analysis has also helped identifying cropping pattern, extent of unauthorized irrigation and poor recovery of water rates, thus, providing inputs for changes in policies and operational plans. Furthermore, spatial analysis of crop sowing periods and crop condition assessment have thrown up policy issues of relevance to irrigation scheduling, canal maintenance and agricultural productivity. Apart from performance evaluation of irrigation systems, multi temporal satellite data have also been used to map current status and to monitor the spatial extent of water logging and soil salinity and / or alkalinity through the years in most of the irrigation projects. Such exercise has also helped in evaluation of the progress and effectiveness of reclamation programmes by monitoring the extent and magnitude of the problem.

Reservoir Sedimentation

Many reservoirs built at a huge investment are undergoing rapid silting and loss of storage capacity and consequent reduction in economic life of reservoir. The analysis of sedimentation data of Indian reservoirs show that the annual silt rate has been generally 1.5 to 3 times more than the designed rate and the reservoirs are generally losing capacity at the rate of 0.30 to 0.92 per cent annually (Dhruvanarayan and Rambabu, 1983). Conventional hydrographic surveys to reassess reservoir capacity are both costly and time-consuming. Multi temporal satellite data have been used as an aid to capacity survey of many reservoirs. In addition, space-borne multi-spectral measurements have also been related to suspended sediment load in many reservoirs, which provide information on the sediment distribution, circulation pattern and active silting zones. A comparison of turbidity levels in irrigation tanks derived from satellite data can, thus, assist in initiating de-silting operations or catchment area treatment.

Watershed Management

Inappropriate land use practices in the catchment area leads to accelerated soil erosion and consequent silting up of reservoirs. Watershed management is thus an integral part of any water resource project. Since the treatment of a large number of watersheds is not feasible simultaneously, watersheds requiring immediate attention are prioritized. The prioritization of watershed is based on sediment yield potential so that the treatment would result in minimizing sediment load into the reservoir. Satellite data have been extensively used in many watersheds, for deriving the parameters of the Sediment yield Index (SYI) model developed by all India Soil and Land Use Survey, Ministry of Agriculture, Government of India. Sediment yield prediction models have been used to provide quantitative silt load estimate in watershed.

Space-borne multi-spectral data have been used to generate baseline information on various natural resources, namely, soils, forest cover, surface water, ground water and land use / land cover and integration of such information along with slope and socio-economic data in a Geographic Information System (GIS) to generate local-specific prescription for sustainable development of land and water resources development on a watershed basis. The study covering around 84 M. ha and spread over 175 districts has been taken up by the Department of Space, Government of India under a national level project titled 'Integrated Mission for Sustainable Development (IMSD)'. Implementation of appropriate rainwater harvesting structures in selected watershed under this programme has demonstrated the significant benefits by observing rise in ground water levels through increased ground water recharge and agricultural development of what was once a barren area. Multi-year satellite data is also used to monitor the impact of the implementation of watershed management programmes.

Flood Management

Floods have been causing severe damages to agriculture, human settlements and have been claiming several human lives throughout the history. India is the worst flood hit country in the world after Bangladesh. India accounts for one-fifth of global death count due to floods. About 40 million hectares or nearly one-eighth of the country is flood-prone and one-fifth of this area experiences floods in any one year. The chronic flood prone river basins in India are the Ganges and the Brahmaputra. These Himalayan rivers flowing down the hills cause flood problems in the states of U.P., Bihar, West Bengal and Assam due to high discharges concentrated during monsoon months (June to September) and large volumes of sediment is carried to the plains.

Space-borne multi-spectral measurements have been used for mapping flooded areas and the assessment of flood damages by the National Remote Sensing Agency since1970s. However, the methodology for flood mapping has become operational only in 1986. Soon after the occurrence of a significant flood event, the spatial extent of inundated area is mapped and damage statistics generated and sent to concerned State and Central Government authorities for providing relief. Active microwave sensor data with cloud penetration capability have been used since 1993 for flood mapping and damage assessment.

Geographic information System (GIS) has been used to generate Digital Elevation Models (DEM) of a flood prone area in one of the southern states (Andhra Pradesh) with the help of in situ measurements made by Global positioning System (GPS) to enable assessment of spatial inundation at different water levels in the river. When the satellite derived land cover/ use and ancillary ground-based socio-economic data is draped over DEM, flood vulnerability can be assessed to provide locale-specific flood warning.

Drought Management

Drought is one of the worst natural disasters affecting the social and economic life of millions of people every year. Developing countries have much large losses from drought compared to developed countries. The World Meteorological Organization estimated that in the quarter of a century since 1967 to 1991 droughts have affected 50% of the 2.8 billion people who suffered from weather related disasters. Moreover, 1.3 million of the 3.5 million people killed were due to direct or indirect causes of drought (Obasi, 1994), with the highest loss of about 8.16 million dead and 1314 millions affected respectively in Asia and 2.07 and 245 million respectively in Africa. The study indicated that more than 500 million people live in the drought prone areas of the world and 30% of the entire continental surface is affected by droughts or desertification process.

Owing to the abnormalities in monsoon precipitation in terms of both spatial and temporal distribution, drought is a frequent phenomenon over many parts of India. Out of net sown area of 140 million hectares about 68% of the area is reported to be vulnerable for drought conditions and about 50% of the drought prone area is classified as severe where frequency of drought is regular. Based on coefficients of variation of rainfall derived from historic data, occurrence of drought is a reality once in 3-4 years in major part of the country (Rao, 1999).

Timely and reliable information about the onset of drought, its extent, intensity, duration and impacts can limit the drought related losses of life, minimize the human suffering and reduce damage to the economy and environment. Space technology has made substantial contributions in every aspect of drought management such as preparedness, early warning, monitoring and mitigation. Remote sensing data from geo-stationary and polar orbiting weather satellites such as INSAT, NOAA, METSAT and other global data is used as major inputs to all the three types rainfall predictions, such as long term seasonal predictions, medium range predictions and short term predictions, while communication satellites have significant potential for real-time dissemination of information. Normalized Difference Vegetation Index (NDVI) derived from NOAA AVHRR (1.1km) and/or IRS WiFS (188m) imagery is now being continuously used to monitor drought conditions on a real time basis often helping the decision makers to initiate appropriate strategies for recovery by changing cropping pattern and practices.

Water Quality

Nearly one million children in India die of diarrhoeal diseases each year directly as a result of drinking unsafe water and living in unhygienic conditions, some 45 million people are affected by water quality problems caused by pollution, by excess fluoride, arsenic, iron or by the ingress of salt water. Millions of people do not have adequate quantities of safe water, particularly during summer months. Increasing environmental concerns on deteriorating water quality are not well supported by ground monitoring mechanisms. Remote sensing of water quality can complement ground efforts in mapping and monitoring point and non-point pollution sources, the influx and dispersal of pollutants in the aquatic environment and consequent impact such as algal blooms and weed growth (UNICEF, 1998). Point source identification calls for high spatial resolution satellite data. Regional models of non-point source pollution loading will be benefited from remote sensing inputs on land use-land cover, supported by sample ground data collection. Growth of aquatic weeds and algal blooms, an indication of eutrophication, have been mapped from satellite and aircraft data. Ground truth requirements are also more stringent than in land remote sensing. The GIS technology provides enhanced capability for water quality modeling.

Ground Water

Nearly half of the current irrigation potential of 90 million ha and large urban and rural population draw water from ground water resources. Studies have shown how integration of geological, geophysical and remote sensing data in geo-statistical models can help improve the success rates of high yielding irrigation wells.

Assessment of Ground Water Resource

In many parts of the country, ground water levels have started declining and in some parts, the existing wells have dried up due to lowering of water table. In the recent years satellite data have proved to be highly useful in mapping and monitoring the ground water in over exploited areas in conjunction with the ground hydro-geological information and providing additional inputs for more realistic estimation of ground water resource. A modified approach has been evolved by Reddy (1999) by integrating lithology, landform, structure, slope, and soil and land use/ land cover information for estimating the individual unit wise ground water recharge. This approach not only helps in reducing the error of averaging of aquifer parameters over large areas but also facilitates more systematic planning and management of ground water resource. As irrigation accounts for more than 90% of ground water utilization in the rural areas, irrigated area statistics form one of the key inputs for ground water draft estimation. Satellite data helps in accurate identification, mapping of the area irrigated by ground water i.e. other than canal and tank commands. Thus, it helps in understanding approximately the amount of ground water that is being used during different seasons of the year in different zones.

Preparation of Ground Water Prospect Maps

In India, more than 90% of rural and nearly 30% of urban population depends on ground water for meeting their drinking and domestic requirements. In addition, it accounts for nearly 60% of the irrigation potential created in the country. The distribution of ground water is not uniform in all the regions. The spatial and temporal variations in rainfall and regional / local differences in geology and geomorphology have led to an uneven distribution of ground water in different regions across the country. This uneven distribution (poor prospects) and indiscriminate tapping (over-exploitation) in certain zones are the main reasons leading to scarcity of drinking water in many parts of the country. In view of this, a large number of habitations in the country have remained as problem villages not having sustainable drinking water sources. In this context, the ground water source finding for planning sustainable drinking water schemes assumes great significance.

In the past, the ground water studies were conducted by conventional hydro-geological surveys. In this approach the geological and hydrological information was collected in the field along traverse lines and representative points. With the development of remote sensing technology, the mapping procedures have undergone significant change.

Being a hidden resource, ground water is not directly amenable to remote sensing. However, remote sensing data provide integrated information on several factors, which directly or indirectly control the movement and occurrence of water below the ground. The synoptic view of the satellite data helps in mapping different lithologic, structural and morphological units in their correct spatial relationship, and the multi-spectral data provide certain additional information which are not otherwise easily observable on the ground.

The Concept

Ground water regime is a dynamic system wherein the water absorbed at the earth's surface passes through geological strata and recycled back to the surface. In this process, apart from geological framework, various elements like relief, slope, ruggedness, depth of weathering, nature of weathered material, thickness of deposition and nature of deposited material, distribution of surface water bodies, river / stream network, canal commands, the extent and intensity of precipitation etc. also exercise significant control on the ground water regime. Thus, the framework in which the ground water occurs is as varied as that of rock types, as intricate as their structural deformation and geomorphic history, and as complex as that of the balance amongst the lithological, structural, geomorphic and hydrologic parameters. The possible combinations and their intricacy are virtually infinite leading to the unavoidable conclusion that the ground water conditions at a given site are unique and are not completely amenable to scientific understanding. Some of the conditions are often obscured and are not clear even during the field observations. However, factor-wise analysis, systematic mapping, data integration and interpretation based on conceptual understanding are helpful in overcoming this problem to some extent.

Factors Controlling the Ground Water Regime

Varying degrees of uncertainly and inconsistency are inherent in the existing methodology (conventional hydro-geological mapping). Hence, a new approach which is systematic and simpler with well defined units has been evolved. In this approach, all the variables controlling ground water regime have been grouped into the following four parameters:

1. Geology / Lithology
2. Geomorphology / Landforms
3. Geological Structures
4. Recharge Conditions

If the information on these four parameters are precisely known, it is possible to understand the ground water regime better, and visualize the gross aquifer conditions, like nature of the aquifer material, type of aquifer, the type of wells suitable, the depth range, yield range, success rate, area of influence, sustainability in each unit.

The influence of each of these parameters on the ground water regime and the information that can be derived in respect of these four parameters using the satellite imagery are briefly discussed.

Geological Mapping

Remote sensing provides the basis for discrimination and differentiation of rock types. Though, direct identification is limited to a few contrasting rock types, many of the rocks can be discriminated based on their spectral and morphological characteristics (Table 5.1).

Once, rock types are identified, the physical continuity of individual rock units can be easily traced and the exact shape, size and geometry can be identified and mapped with minimum ground survey and more accurately by using satellite imagery.

Mapping Geological Structures

The most obvious structural features that are important from the point of groundwater are lineaments. They are seen on the satellite imagery as linear alignments of structures, lithology, topography, vegetation or drainage anomalies, either as straight lines or curvilinear features. They can be further classified into faults, fractures, joints, bedding traces, contacts of rocks, shear zones etc. based on the image characteristics and association, after corroborating with the ground information. In hard rock areas, faults and fractures mainly act as conduits for groundwater movement and form prospective groundwater zones.

Table. 5.1 The morphological and spectral characteristics of different rock groups as expressed on the satellite imagery (Source: P.R. Reddy, 1987)

S.no.	Rock Type	Physical Characteristic
1.	PLUTONIC IGNEOUS ROCKS (e.g. Granite, Syenite, Diorite, Gabbro, Dunite, Pyroxenite Amphibolite, etc.)	Occurs as positive landforms, boldly standing-out as hills and inselbergs, in the form of exfoliation domes, monadnacks, bornhardts, tors, koppies etc. with radial, concentric, annular and mixed complex drainage patterns and sparse vegetation.
2.	DYKES AND OTHER INTRUSIVES (e.g. Dolerite/Peridotie dykes and sills, Pegmatite bodies, Quartz reefs)	Display, linear trends and appear as wall like features, resistant ridges, hog-back ridges etc. If the surrounding rocks are more resistant, they form linear depressions and furrows.
3.	VOLCANIC ROCKS (e.g. Basalt, Andesite, Dacite, Rhyolite)	Volcanic cones, lava plains and tongues with branching pattern, flow lines, pyro-clastics, ropy structures, et., are common. They show crude stratification, with successive lava flows varying in thickness, composition, and sometimes interbedded with tuff and ash beds. Occur, as mesas, buttes and flat valley-floors with mixed complicated drainage pattern due to columnar jointing.
4.	SEDIMENTARY ROCKS (e.g. Sandstone, Quartzite, shale, limestone, conglomerate)	Horizontal to sub-horizontal beds, often with contrasting lithologies, forming, hogback homoclinal ridges, cuestas and stratiplains. Ridge and valley topography with modified trellis and parallel drainage patterns. Stratification, lensing, etc., can be seen on large scale aerial photographs. Arenaceous rocks have light tone and low drainage density. Argillaceous rocks have medium tone and high drainage density.

Table. 5.1 *Contd.....*

Sl.	Rock Type	Physical Characteristic
5.	METAMORHIC ROCKS (e.g. Quartzite, phyllite, schist, gneiss)	Appear, as elongate parallel belts of valleys and hills, often display bedding, foliation, schistosity as thin parallel lineations; folded, crenulated, wavy or crinky pattern, characterised by trellis, rectangular and parallel drainage patterns; quartzites usually occur as strike ridges.
6	CALCAREOUS ROCKS (e.g. Limestone, Dolomite, marble, gypsum, chalk)	Indicated, by sinkholes, solution-cavities, depressions, blind-valleys etc, and karst topography, often marked by pitted appearance. Deranged type of drainage pattern is common.
7.	DURICRUSTS (e.g. Laterite and Bauxite)	Occur, as plateaux, with bright tone on the imagery with dendritic drainage and sparse vegetation.

***N.B.** : Based on their spectral characteristics manifested on the imagery as different tones (colours), further differentiation within the above rock groups can be made into acidic rocks (light tone), intermediate rocks (medium tone), basic and ultrabasic rocks (dark tones), etc. The moist and thin soil covered areas appear as dark toned areas on the image. Vegetation appear as bright red patches in the standard F.C.C. images and water bodies appear as light blue to black depending on their depth and turbidity. Generally, the morphology (landforms) indicate the textural and structural variations and tone on the image indicates the compositional variations in the rock types. By combining both morphological and spectral (tonal) characteristics, different rock groups / types can be mapped.*

Geomorphological Mapping

Geomorphology exercises a significant control over groundwater regime. The relief, slope, depth of weathering, type of weathered material, thickness of deposition, nature of the deposited material and the assemblage of different landforms play an important role in defining the ground water regime, especially in the hard rocks and the unconsolidated sediments. Satellite imagery, due to its synoptic view, facilitates better appreciation of geomorphology and the helps in mapping of different land-forms and understanding their origin, sequence of evolution, material content and other characteristics. Palaeo-river course, buried channel, alluvial plain, bajada deposit, closed valley, deeply weathered pediplain, glacial till, fracture / fault line valley, karst landscape, which form prospective ground water zones, are most evident on satellite data.

Hydrologic Information and Recharge to Groundwater

Recharge is the most important factor in evaluating the groundwater prospects. Recharge to groundwater depends on precipitation, surface water bodies like reservoirs, lakes, tanks, streams, canals, canals commands, favourable rocks, landforms and structures etc.

Satellite imagery provides information on surface water bodies, canals, canal commands, and irrigated fields, which directly contribute to groundwater recharge. Based on geomorphic and hydrologic analysis, the areas can be classified into runoff, recharge, storage and discharge zones, and the amount of recharge to groundwater from different sources can be estimated.

Classification System

For mapping the four parameters mentioned before, i.e. lithology (geology), landforms (geomorphology), geological structures and recharge conditions, well defined classification systems have been evolved as furnished in Tables 5.2, 5.3, 5.4 and 5.5 respectively.

Based on the visual interpretation of satellite imagery in conjunction with the existing geological / hydro-geological data and limited field checks, all the four parameters have to be separately mapped. The rock (lithologic) units have to be indicated with numerical code numbers, landforms with alphabetic codes, geological structures and hydrological data with line symbols.

Table 5.2 Hydrogeological (lithological) classification system (Source : P.R. Reddy, 1991)

Code No.	Rock Group	Code No.	Lithologic Unit
1.	UNCONSOLIDATED SEDMENTS	11. 12. 13. 14. 15. 16. 17.	Gravel, Sand, Silt Clayey Sand Sandy clay Clay Layers of sand, silt & clay Colluvium Others
2.	RESIDUAL CAPPINGS	21. 22. 23. 24. 25. 26.	Laterite (Ferricrete) Bauxite (Alucrete) Kankar (Calcrete) Chert (Silcrete) Secondary / Detrital laterite Others
3.	DECCAN TRAPS & INTER-TRAPEANS	31. 32. 33. 34. 35. 36. 37.	Massive Basalt Vesicular Basalt Tuffaceous Basalt Rhyolite Red/Green Bole Intertrappeans Others
4.	OTHER VOLCANICS/ METAVOLCANICS	41. 42. 43. 44. 45.	Basalt Rhyolite Dacite Andesite Others
5.	SEMI-CONSOLIDATED SEDIMENTS	51. 52. 53. 54. 55. 56. 57. 58.	Sandstone & conglomerate Shaly Sandstone Sandy Shale Shale, Clay, Coal / Lignite Shell-limestone Sandstone with shale/coal partings Shale with sandstone partings Others

Code No.	Rock Group	Code No.	Lithologic Unit
6.	CONSOLIDATED SEDMENTS	61.	Shaly Limestone
		62.	Thin Bedded Limestone
		63.	Shale with limestone bands/lenses
		64.	Thick bedded Limestone
		65.	Compact Shale
		66.	Thick bedded Sandstone / Quartzite
		67.	Thin Bedded Sandstone / Quartzite
		68.	Others
7.	INTRUSIVE ROCKS	71.	Quartz Reef / Quartzite
		72.	Basic Dyke / Pipe Rock
		73.	Basic Sill
		74.	Quartz / Pegmatite / Aplite intrusion
		75.	Others
8.	CRYSTALLINE ROCKS	81.	Massive Granite & Plutonic Rocks
		82.	Granite and Gneissic Complex
		83.	Phyllite / Schist / Slate with bands of Quartzite / Quartz reefs
		84.	Quartzite with Shale / Phyllite bands
		85.	Phyllite, Schist and Slate
		86.	Gneiss
		87.	Schist and Gneiss mixed
		88.	Meta-Basics / Metavolcanics
		89.	Others

Additional lithological units which are not covered in the above classification may be added in each group (dashed lines indicate scope for adding additional units)

Table 5.3 Hydrogeomorphological (landform) classification system (Source: P.R. Reddy, 1991)

Physiography	Code	Geomorphic Unit / Landform
HILLS	**CH**	**COMPOSITE HILLS**
	CHR	Ridge type
	CHR	Homoclinal type
	CHC	Cuesta type
	CHB	Butte / Mesa type
	CHD	Dome type
	CHM	Massive type
	RH	**RESIDUAL HILLS**
	RHR	Ridge type
	RHH	Homoclinal type
	RHC	Cuesta type
	RHB	Butte / Mesa type
	RHD	Dome type
	RHM	Massive type
	-	**INSELBERGS**
	LR	Linear Ridge
	FR	Folded / Accurate Ridge
	C	Cuesta
	M	Mesa / Butte
	D	Dome
	I	Inselberg
	PIC	**PEDIMENT - INSELBERG**
	PRC	Ridge Complex
	PCC	Cuesta Complex
	PMC	Mesa / Butte Complex
PLATEAUS	**UP**	**UPPER PLATEAU**
	UPS	Undissected
	UPD	Dissected
	MP	**MODDLE PLATEAU**
	MPU	Undissected
	MPD	Dissected

Table 5.3 *Contd....*

	LP	**LOWER PLEATEAU**
	LPU	Undissected
	LPD	Dissected
	OPF	**OUTER FRANGE OF PLATEAU**
PIEDMONT ZONE	-	**EROSIONAL**
	PS	Piedmont Slope
	-	DEPOSITIONAL
	PA	Piedmont Alluvial Plain
	PF	Piedmont Alluvial Fan
PLAINS	-	**EROSIONAL**
	PD	Pediment
	PP	Pediplain
	PPS	Shallow wealthered
	PPM	Moderately wealthered
	PPD	Deeply weathered
	PPG	Gullied
	PPL	Lateritic
	SP	**STRIPPED PLAIN**
		Shallow basement
		Moderate basement
	AP	**ALLUVIAL PLAIN**
	APS	Shallow
	APM	Moderate
	APD	Deep
	APG	Gullied
	FP	**FLOOD PLAIN**
	FPS	Shallow
	FPM	Moderate
	FDP	Deep
	FPG	Gullied
	DP	**DELTAIC PLAIN**
	DPS	Shallow
	SPM	Moderate
	DPD	Deep
	DPG	Gullied

Table 5.3 *Contd....*

	CP	**COASTAL PLAIN**
	CPS	Shallow
	CPM	Moderate
	CPD	Deep
	CPG	Gullied
	EP	**EOLIAN PLAIN**
	EPS	Shallow
	EPM	Moderate
	EPD	Deep
VALLEYS	-	**EROSIONAL**
	SV	Structural Valley
	CV	Closed Valley
		DEPOSITIONAL
		Filled-in-valley
		Intermontane Valley
OTHER	-	**FLUVIAL LANDFORMS**
LANDFORMS	BC	Buried Channel
	MC	Migrated Channel
	PB	Point Bar
	CB	Channel Bar
	NL	Natural Levee
		EOLIAN LANDFORMS
	DF	Dune-Field
	DC	Dune-Complex
	PY	Playa
	AR	Absorbed River/Stream
	-	COASTAL LANDFORMS
	B	Beach
	BR	Beach-Ridge
	BRC	Beach-Ridge Complex
	PBC	Plain-beach Ridge Complex
	SW	Swale

Additional landforms if any depending upon the areas may be added in each group

Table 5.4 Classification of geological structures (P. R. Reddy 1991)

Structure	Category	Field / Observed
1. Bedding		
	Gentle (<15°)	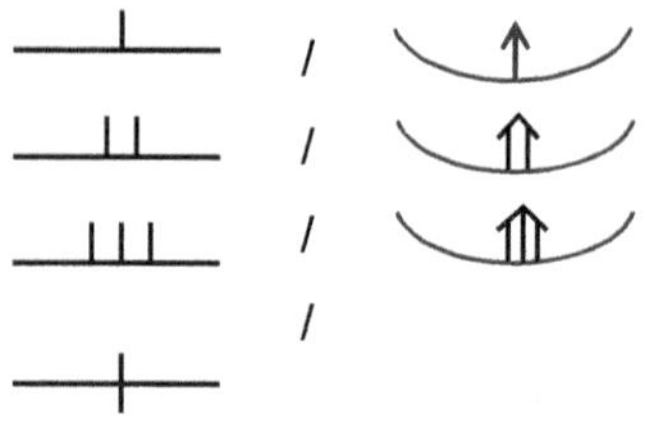 /
	Moderate (15-45°)	/
	Steep (45-85°)	/
	Sub-vertical to	/
	Vertical (>85°)	
2. Schistosity / Foliation		
	Moderate (<45°)	/
	Steep (45-85°)	/
	Vertical (>85°)	/
3. Faults		
	Confirmed	F———F
	Inferred	F- - - - - -F
4. Fractures / Lineaments		
	Confirmed	
	Inferred	
5. Shear Zones		
	Confirmed	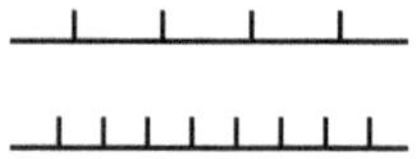
	Inferred	
6. Folds		
	Anticline/Antiform	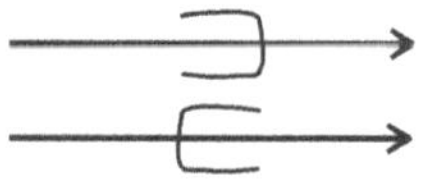
	Syncline/Synform	
7. Trend Line		
	Bedding/Foliation etc.	

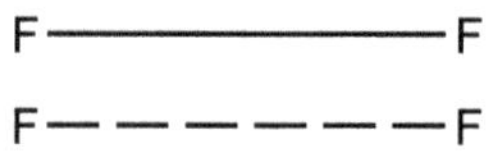

Table 5.5 Classification of recharge conditions (P.R.Reddy, 1991)

Excellent	Recharge from continuously irrigated canal commands, permanent water bodies, perennial streams etc. (Continuous recharge throughout the year with shallow water tables conditions)
Very good	Recharge from temporarily / seasonally irrigated canal commands, seasonal / ephemeral water bodies / streams (High recharge for part of the year)
Good	Large recharge area with high rainfall, facilitating good recharge.
Moderate	Limited recharge area and moderate rainfall or seepage from other sources or limited by other factors
Limited	High relief areas occupying large areas with heavy runoff, very limited recharge only along narrow valleys.
Poor	Very limited recharge area / poor rainfall / shadowed by ground water barriers like linear ridges, dolerite dykes etc.
* The recharge condition for each unit may be generalised and indicated in the legend	

Preparation of Map

The ground water prospect map is prepared by integrating the information on the four parameters mentioned before. For this purpose, by combining the rock type and landform information, integrated rock-cum-landform units (hydro-geomorphic units) are evolved. These hydro-geomorphic units will have less heterogeneity and exercise better control on the ground water regime. While mapping these hydro-geomorphic units, the boundaries between rock type and landform have to be made co-terminus wherever, within a rock unit two or more landforms and vice-versa are possible. The structural information as per Table 5.4 has to be represented on the map with appropriate line symbols. Similarly, the hydrological information like depth to water table, rainfall, observation wells with their yield range etc. have to be represented on the map with contours and point symbols in different colours. Thus, the ground water prospective zones map is prepared which provides much better information than the conventional hydro-geological map.

Legend and Colour Scheme

An innovative legend has been developed for preparing the ground water prospective zones map. In this legend, individual hydro-geomorphic unit-wise rock type, landform and recharge conditions are listed out in tabular format and taking these factors into consideration the ground water prospects of each unit have been evaluated in terms of (i) nature of the aquifer material, (ii) type of aquifer, (iii) type of wells suitable, (iv) depth range of wells, (v) yield range of wells, (vi) success rate of wells etc. Further, to make the hydro-geomorphic units less heterogeneous, the faults, fractures and lineaments occurring in different hydro-geomorphic units can also be separated and mapped as fracture zones and the ground water prospect for these fracture zones can be indicated separately. Thus, the heterogeneity arising out of fracture system in hard rocks can be reduced to a great extent.

In the maps, each hydro-geomorphic unit's ground water prospects are indicated by different colour. For this purpose, a colour scheme has been devised using 7 colours of VIBGYOR colour spectrum for representing seven categories of ground water prospects i.e. ***excellent, very good, good, moderate, poor, limited*** *and* ***nil***, based mainly on the yield range of wells in different hydro-geomorphic units. The excellent prospects can be indicated with violet colour followed by ***very good*-indigo, *good*-blue,**

***moderate*-green, *poor*-yellow**, ***limited*-orange** and **nil**-with **red** colours. Wherever, the quality of water is not good / not potable it is indicated with **gray** colour. Based on this colour scheme it is easier for the user to identify the target zone around a specific point using the ground water prospect map.

Projected Scenario

As mentioned earlier, sustainable water resources management calls for optimal utilization of available water resources to meet the domestic, agricultural and industrial demands. A thorough understanding of hydrological phenomena and fresh water availability and its utilization is a pre-requisite to achieve the goal.

Hydrological Phenomenon

Identification and quantification of key hydrological variables across a range of scales, and the development or modification of hydrologic process need to be taken up for furthering our understanding in hydrological phenomena. Furthermore, estimation of water balance at the land surface for which precise information on precipitation, runoff, evapo-transpiration (ET), soil water and ground water storage is also required. Currently only point observations of precipitation are available. The Earth Observing System (EOS) sensors, namely Tropical Rainfall Mapping Mission (TRMM) – Microwave Imager (TMI), Visible Infrared Scanner (VIRS), Advanced microwave Sounding Unit (AMSU) and Advanced Microwave Scanning Radiometer (AMSR) may provide data relating to precipitation and clouds. In addition, Atmospheric Infrared Sounder (AIRS) and microwave sounders (AMSU and HSB – Humidity Sounder Brazil) will together provide information regarding precipitation volumes, cloud thickness and cloud water content (Anonymous, 1999).

The near surface soil moisture has been estimated through passive radiometry and active radar. It has, however, met with limited success. Direct satellite observation of near surface soil moisture is currently under experimental study. Conventional and experimental (such as inflatable antenna) L-band radiometer under NASA's Small Satellites and New Millennium, development of algorithm and multi-sensor signals, and improved representation of soil moisture in Land Surface Models of sub-humid regions across a variety of scales may help mapping soil moisture.

Conclusions

Water resource management need to be planned and implemented within the framework of integrated resource management which requires consideration of a range of impacts, sometimes extending far beyond the immediate hydrological system, and over considerable time periods. Space borne multi spectral measurements have in some cases replaced ground based observations and in others complemented at varying levels. Improved spatial, spectral and temporal resolution data from planned missions, namely IRS-P5 (Cartosat-1) / IRS-P6 (RESOURCESAT), ORBIMAGE, Resource-21 and currently operating Ikonos-2 will lead to a quantum jump in our capabilities for remote sensing application towards water resources management. In complement to these developments in the space segments, dramatic improvements in data processing, analysis, interpretation and integration would help in improving the strategies for water resources management.

References

Annoymous, 1999, EOS Science Plan: *The state of the Science in the EOS Programme*, In M.D. Kind, R. Greenstone, W. Bandeen (Eds.), NRSA publication no. NP-1998-12-069-GSFC.

Dhruva Narayan, V.V. and Ram Babu, 1983, Estimation of soil erosion in India, *Journal of drainage division*, ASCE, v. 109(4), pp 419-434.

FAO, 1990, *An international action programme on water and sustainable agricultural development: a strategy for the implementation of the Mar del Plata Action Plan for the 1990,* Food and Agriculture Organization, of the United Nations, Rome.

Kasturirangan, K., 1996, *A Space Agenda for Water: Challenges and Perspectives*, Sixth Vikram Sarabhai Memorial Lecture delivered during National Symposium on Remote Sensing for Natural Resources with special emphasis on Water Management, Dec.4, 1996, ISOR-HQ-SP-80-96.

Mohile, A.D., T.S. Bathija, V.K. Jyothi, A.D, Bharadwaj, R.S. Pathak, and M. Sivadas, 1996, Inter Basin Transfer of Water for National Development Problem and prospects, *Journal of Indian water Resources society*, vol.2no.2, April, 1996, pp 1-26.

Obasi, G.O.P., 1994, WMO's role in the International decade for Natural disaster reduction, *Bulletin of the American meterological society*, Vol. 75, pp 1655-1661.

Rao, D.P., 1999, *Space and Drought Management in Asia – Pacific region*. Space Forum, v. 4, pp.223-247.

Reddy, P.R, 1987. *Geological and geomorphological studies through remote sensing* . Proc. Workshop on Remote Sensing Technologies, Malaysia, 21-26 June

Reddy, .P.R. 1991. *New concepts and approach for ground water evaluation with special reference to remote sensing*, Ph.D thesis, Osmania University, Hyderabad, India (unpublished)

Reddy., P.R, 1999. Remote sensing in ground water studies, In '*Remote Sensing for Earth resources*' (ed. D.P.Rao) , AEG, Hyderabad

Shiklomanov, I. A., 1998, *Water Use in the World, Present situation and future needs*, prepared in the frame work of International Hydrological Program, published by UNESCO.

UNICEF, 1998, *Emerging fresh Water Crises in India.*

WHO/UNEP, 1991, *Water Quality: Progress in the implementation of the Mar del Plata Action Plan for the 1990's.* WHO, Geneva, and UNEP, Nairobi.

WMO/UNESCO, 1991, *Water Resources Assessment: Progress in the Implementation of the Mar del Plata Action Plan for the 1990s*, WMO,Geneva, and UNESCO, Paris.

World Resources Institute, 1992, *World Resources 1992-93*, Oxford University Press, New York.

6

GIS and Remote Sensing in Watershed Management

I.V. Murali Krishna

Centre for spatial information technology, Jawaharlal Nehru Technological University, Hyderabad 500 028, A.P., India

(e-mail iyyanki@icord.org)

Key words : Remote sensing, GIS packages, GIS layer creation, soil cover complex, Veligonda Irrigation Project.

Introduction

Remote sensing and GIS, grouped under an emerging tool of technology called geospatial information technology, are the most valuable for watershed management to improve the quality of life and deal effectively with various engineering tasks concerned with earth, environment and people. The words geospatial and spatial are being commonly used and they essentially mean same and refer to geographical information which is synonymous with any information, that has a geo-reference. It is about the geographical distribution of any land use/ land cover feature, terrain feature, communication network or a conventional utility feature. It has a geographical component as well as attribute (non-spatial) information. The geospatial information technology covers a wide range of topics, disciplines, subjects and technological tools. These are concerned with surveying, geographical information systems (GIS), Global positioning systems (GPS), computer aided drafting (CAD), cartography, transportation management, construction management, water supply and sanitation, photogrammetry, remote sensing, terrain mapping and visualization, surface and ground water resources management, urban and rural planning, coastal zone management, data base management, software engineering, web technology, computer graphics, pattern recognition, expert systems and image processing (Burrough, 2002; Gregory and Walling, 1973 ; Pimentel, 1993).

The Indian space program has provided tremendous impetus for the growth of the geospatial information technology in India. The very high-resolution data from the IRS (Indian Remote Sensing Satellite) LISS III and Pan (Panchromatic) sensors has given scope for scientists and engineers to use the remote sensing data along with tools of information technology. The tasks are related to application of remote sensing, photogrammetry and topographical surveying and their integration. Georeferencing, image registration and merging, orthophoto generation, image map generation are some of the very significant information oriented data product generation tasks, that are of immense use in planning, design, construction, management, and monitoring.

This chapter is concerned with application of GIS for watershed management and as such let us briefly reviews few aspects of the watershed for assessment of the potential and scope of application of the technology. There are different approaches in watershed management like problem-oriented approach, three dimensional approaches and integrated approach. Any discussion of these details is beyond the scope of this chapter as the intention here is to highlight the applications of GIS for watershed management

Watershed Management

Watershed can be defined as the area drained by a stream or a system of streams such that all the surface runoff originating in this area leaves the area in a concentrated flow through a single outlet. A watershed, which is a manageable independent hydrological unit, can be taken as the basic unit of development planning. In Hydrological terms, watershed is an independent unit where runoff from the area drains off from a common outlet. The surface runoff follows the general slope and comes out through a single point. Watershed gives idea about the available water in an area. Another advantage with watershed as a unit is that, even a geographically large watershed in a much-damaged state can also be rectified part-by-part dividing watershed into sub-watersheds. Watershed is the ideal unit for management of the natural resources and for sustainable development and sustainable life styles. The concept of watershed as a basic unit does not mean that the administrative boundaries should be on watershed basis, but it calls for planning natural resource management on a watershed basis. The implementation, ideally could be on watershed basis. Every watershed is having its own characteristic features and problems, thus making it unique. In order to derive maximum benefit out of these resources, in a sustainable manner, it is preferable to treat the watershed as a single unit.. All watersheds can be divided into smaller sub-watersheds. As each watershed or sub-watershed is an independent hydrological unit, any modification of the land used in the watershed or sub-watershed will reflect on the runoff as well as sediment yield of the watershed. The main objective of the watershed management is proper utilization of land as per its capacity and limitations and prevent soil erosion and design methods for groundwater recharging. This helps in maximizing productivity per unit area of land under the given constraints with integrated approach and better farming systems.

Remote Sensing

The first near operational remote sensing data was available from the US satellite Landsat in 1972 with a spatial resolution of about 80 meters in multi spectral mode. The Indian Remote sensing satellite IRS 1A was launched in 1988 with resolutions 79 and 36 meters. Later IRS 1B is launched to maintain the continuity of satellite data. This was followed with IRS 1C and 1D providing spatial resolution of 23 meters in multi spectral mode and 5.9 meters in panchromatic mode. Added to these, the US based company viz., Space Imaging, launched IKONOS, a very high-resolution satellite. IKONOS provides data at 2.5 meters resolution in multispectral mode and 1 meter resolution in panchromatic mode. The very high resolution data at about 0.6 meters resolution is available from Quick Bird satellite operated by Digital globe. Cartosat, the Indian satellite provides 2.5 meters resolution. India continues to have a pre-eminent position due to its high-resolution satellite development, associated infrastructure, and will continue to be very important in the future. Many water resources applications would be requiring high-resolution ortho photo mosaics. The trend is now set for digital systems by leading photogrammetry companies. This would significantly alter the tasks of topographical surveying and watershed management strategies in terms of time, technology and efficiency. A comparative analysis between conventional and remote sensing study as given by Perumal (1994) is reproduced below.

Table 6.1 A comparative analysis of the cost in rupees per hectare for carrying out integrated resources mapping

Theme	Conventional Survey/ mapping (Rs)	Based on Remote Sensing data (Rs)
Hydro-geomorphology	1.30	0.31
Land use / Land cover	1.00	0.22
Soil	2.00	1.10
Transport Network	0.05	0.05
Drainage	0.05	0.05
Meteorological data	0.01	0.01
Socio-economic data	0.07	0.07
Action plan	0.54	0.54
Total	5.10	2.43

Newer applications of remote sensing are emerging with the availability of high-resolution satellite data processing tools, soft copy photogrammetry and Global positioning system receivers. The high-resolution data and derived products like orthophotos and image maps have gained lot of attention and are very useful for many engineering applications. The requirement of orthophoto and image maps is bound to grow rapidly. This will be driven by the demand for large ortho mosaics for various planning and implementation purposes including the requirement of serving the information on the Internet.

The very high resolution data helps in preparation of very large scale maps. The large scale maps are very useful for irrigation projects planning and management, urban applications, land degradation studies and agriculture inventory etc. For example, identification of existing or potential degraded lands or erosion prone lands, is very much timely to help in planning, reclamation or preventive measures. Methods of erosion detection and assessment using remote sensing techniques are well established based on tone, texture and physiography recognition features. Drainage, precipitation, vegetation, elevation and relief are the factors to be considered in estimating the water-induced erosion. Soil erosion features may occur in a regular sequence of types and intensities along the topographic relief in a certain area. The remote sensing imagery, helps in delineation of large-scale degraded lands.. The image interpretation takes into consideration the direct and indirect features that are identifiable at a given scale. . The remote sensing image helps in identification of erosion features, which are qualitatively apparent, say, as changes in soil type color or vegetation cover etc.

Seasonal changes in vegetation cover and moisture levels, as indicated by tonal changes, can be evaluated to increase the accuracy of interpretation of terrain conditions. In addition any changes occurred in the watershed during the year, such as clear cutting of forest, forest fire, developmental works, provide pictures of changes. Interpretive elements of different terrain features on the multi-spectral satellite images of different available satellite sensor data give a broad view of mapping erosion prone areas. Land capability analysis is a very important component of watershed management. Land capability is an expression of the effect of physical land conditions, including climate, on the total suitability for use without damage for crops that require tillage, for grazing, for woodland and for wildlife. In short, land capability is a measure of suitability of land for use without damage. Remote sensing data when used with conventional data can be an effective tool to arrive at land capability classes. The land capability classification proposed by USDA is an extremely useful method and is being widely used with some local variations. The topics related to soil erosion etc are presented in subsequent sections after the explanation of the principles of GIS.

Geographical Information Systems (GIS)

Geographic Information System (GIS), computer system that records, stores, and analyzes information about the features that make up the earth's surface. A GIS can generate two- or three-dimensional images of an area, showing such natural features as hills and rivers with artificial features such as roads and power lines. Scientists use GIS images as models, making precise measurements, gathering data, and testing ideas with the help of the computer. Many GIS databases consist of sets of information called *layers*. Each layer represents a particular type of geographic data. For example, one layer may include information on the water bodies or cropping pattern in an area. Another layer may contain information on the soil in that area, while another records elevation. The GIS can combine these layers into one image, showing how the water bodies or cropping pattern and elevation relate to one another. Watershed managers might use this image to determine whether a particular part of a water body or bund is more likely to crumble. A GIS database can include as many as 100 layers. A GIS is designed to accept geographic data from a variety of sources, including maps, satellite photographs, and printed text and statistics. GIS sensors can scan some of this data directly-for example, a computer operator may feed a map or photograph into the scanner, and the computer "reads" the information it contains. The GIS converts all geographical data into a digital code, which it arranges in its database. Operators program the GIS to process the information and produce the images or information they need. The applications of a GIS are vast and continue to grow. By using a GIS, scientists can research changes in the environment; engineers can design road systems; electrical companies can manage their complex networks of power lines; governments can track the uses of land; fire and police departments can plan emergency routes. Many private businesses have begun to use a GIS to plan and improve their services.

The Canadian government built the first GIS, the Canada Geographic Information System, during the 1960s to analyze data collected by the Canada Land Inventory. Other governments and university laboratories soon built similar systems. However, GIS systems were not widely used until the late 1970s, when technological improvements and lower costs made computers widely available. GIS sales boomed during the 1980s, as governments and businesses found more uses for the systems. The availability of Remote sensing data and software at an affordable price is prime reason for development of GIS. A number of companies began producing new GIS software to program computer systems to increase their functions. By the early 1990s, about 100,000 GIS systems were in operation.

Geographic Information System (GIS) helps to handle digital maps, analyze them and suggest management strategies. Thematic maps prepared based on Remote sensing data are integrated with non spatial information under a geographic information system. GIS happen to be the prime component of spatial information technology. It is a computerized information system with unique facilities for inputting, analyzing, querying, and managing the spatial and non-spatial data. The bottom line here is the ability for integration of spatial information. Most of the spatial information tools have no facilities for data base development. The conventional data base packages provide access only for data, which is in a tabular form. A comparison of related information systems in various conventional packages like CAD, DBMS, MIS and GIS is given in Fig 6.1 which provides a basis for understanding the extent of respective spatial and non spatial capabilities.

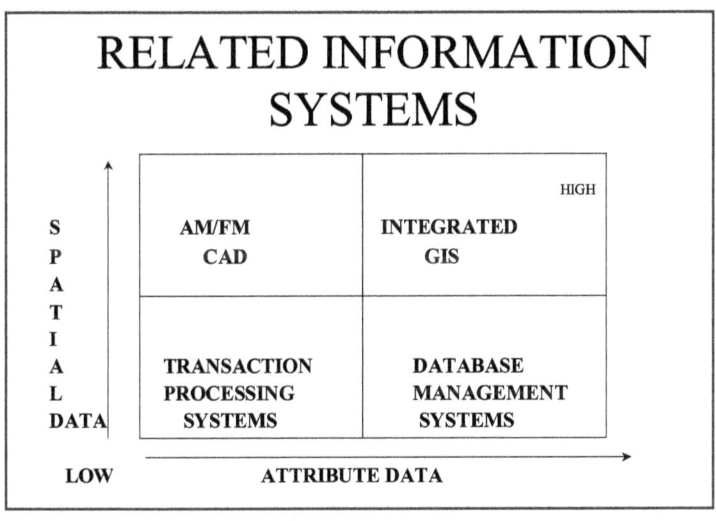

Fig. 6.1 A comparative study of Spatial and attribute data in related Information systems.

In a GIS the spatial analysis and mapping components are very high. The topographic maps prepared by Survey of India on different scales from 1 / 25,000 to 1 / 250,000 are to be used for base map preparation and the thematic information is superimposed on these maps. The remote sensing data facilitate preparation of these thematic maps considering the satellite sensor resolutions as available today. The thematic information, which is in geospatial form is integrated with non spatial or attribute data. This sort of integration is the prime scope and intent of development of any geospatial information system. The concept behind such an integration is shown schematically in Fig. 6.2. A classification of GIS based on its mapping and analysis capabilities can be seen in Fig. 6.3.

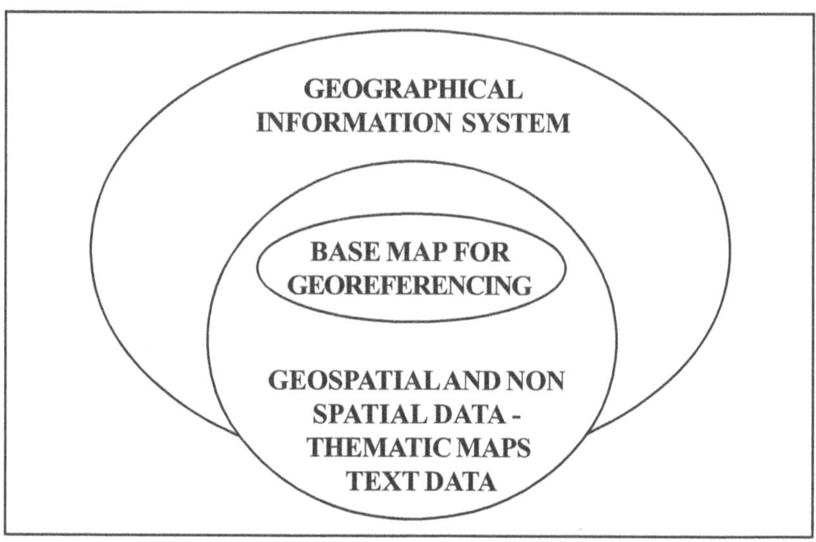

Fig. 6.2 The relationship between GIS and base map/data.

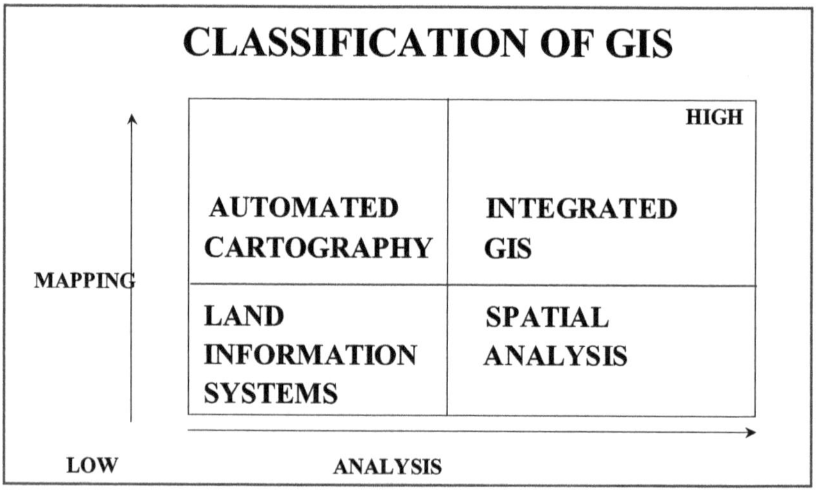

Fig. 6.3 Classification of GIS with reference to mapping and analysis capabilities.

Land information systems (LIS) primarily deal with cadastral data and provide information about land ownership and related details. Automated cartography has a high component of mapping capabilities while spatial analysis could provide only statistical significance of any attribute variation. A very special blend of spatial analysis capabilities for automated cartography leads to development of an integrated GIS. The GIS basically consist of four groups of functions maintenance and analysis of spatial data, maintenance and analysis of non-spatial data and integrated analysis and output generation. The data input, data management, data analysis and data query /output happen to be the prime factors that address the issues under the four groups of functions The scanning and / or digitization are the prime means of data input to a GIS. This involves what is currently called as data conversion or raster to vector conversion (R2V). The remote sensing data in raw or classified mode and photogrammetry form the automatic inputs. The GIS package must have specific analysis capabilities for map overlay and intersection, proximity analysis, network analysis, attribute analysis and merging, map projections, data clipping, data updating and data aggregation. In the context of discussion of salient features of GIS, the basic concept of geographic information system is presented schematically in Fig. 6.4. In the initial stages all applications along with data base are built with GIS occupying a centre position. But the experience of GIS development has shown that GIS along with applications revolve around Data base which occupies central position. In the networking era the web based developments shall occupy a very important place. This suggest that GIS is in a continuously evolving stage.

The developments in information technology have significantly altered the cost scenario with reference to GIS from conception to implementation particularly during the last 5 years. A schematic representation of variation of the cost of various GIS components over the period of late 60s to present time is shown in Fig 6.5.

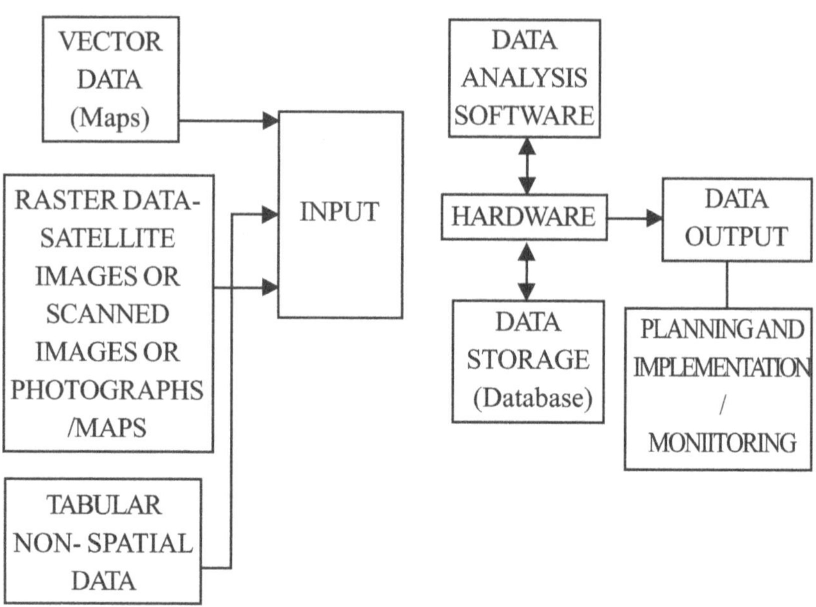

Fig. 6.4 Concept of geographic information system.

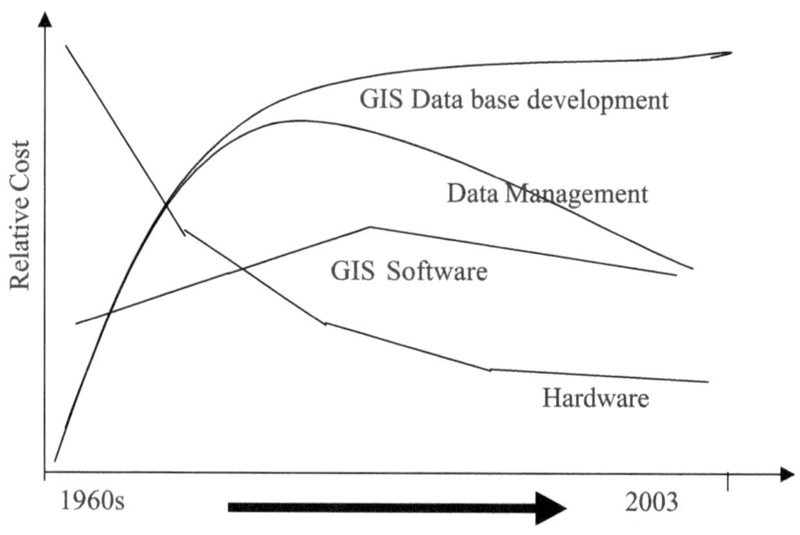

Fig. 6.5 GIS project cost scenario.

The ability for spatial and non-spatial querying is essential component of GIS. The display and reporting capabilities in the form of tables, maps and charts provide the required user interface. The data, which is heart of GIS, must be of high quality. The components of GIS data quality consist of Micro level components, macro level components and usage components Table 6.2 gives details about the elements of these data quality components.

Table 6.2 Data Quality Components

Microlevel components	Macro level components	Usage components
Location accuracy Attribute accuracy Logical Compatibility Logical compatibility Resolution	Completeness Time indirect costs Lineage	Accessibility Direct and

GIS packages

There are a large number of GIS packages available in the international market - mainly from North America, Europe and Australia. Some of the packages are as follows and the list is only illustrative and not exhaustive

ARC/INFO and ARC GIS, PAMAP, SPANS, IDRISSI, ILWIS

Map Info, Microstation , Geo-media, Geoconcept ...

All the packages have proved their utility and depending on the technical requirements and vendor support and funds available, one can decide the package for procurement. As a simple example some features of Arc Info GIS package are given and details of any of the packages can be obtained from user manuals and by browsing the respective websites.

ARC/ INFO (ARC GIS) has been developed by Environmental Systems Research Institute (ESRI), Redlands, Calif., USA (ESRI, 2002). ARC/ INFO is a vector based GIS package, capable of handling both spatial and non - spatial data. It organizes geographical data using vector topological models non - spatial data using relational models in a DBMS. Each vector is either a point feature or a vertex of an arc, each arc is either a line feature or one line for a polygon feature. The arc - node and polygon topology is organized to identify point. line and polygon relations. The cartographic data are then linked to the attribute data through a link - item. ARC/ INFO has a wide range of functionality which has been developed based on a tool - box concept - where each function can be visualized as a tool and having a specific utility. Thus, based on the user requirement, a specific tool or function could be utilized.

Database creation in ARC/INFO is possible through the process of digitization using the ARC Digitizing System (ADS) and the ARCEDIT module. The ADS is a menu - driven module for digitizing and perform editing on spatial features. ARCEDIT is a powerful editing utility having capabilities for feature based editing. These modules include the functions for co ordinate entry using different devices digitizers, screen cursors etc.

Info is a complete relational database manager for the tabular data associated with geographic features in map coverage. ARC/INFO keeps track of and updates map feature attribute table which are stored as INFO data files. INFO can be used to manipulate and update each feature's attribute by performing logical and arithmetic operations on the rows and columns of the table. INFO provides facilities for data definition of data files, use of existing data files data entry & update and sort and query.

ARC/ INFO offers spatial overlay capabilities based on topological overlay concepts. Overlays, Buffer generation, Proximity analysis, Feature Aggregation, Feature Extraction, Transformation, Nearness functions and other integration utilities are available. ARCPLOT module has got capabilities for generating cartographic quality outputs from the database. This includes utilities for interactive map composition, editing map compositions, plotting and printing etc. The map composition functionalities include the incorporation of coverage features as per required scale, generalization, symbolization, transformation etc. Placements of non - coverage features like legends, free text, logos, graphic shapes and so on can also be done.

TIN module of ARC/ INFO can be used to create, store, manage and perform analysis pertaining to the third dimensional data. The modeling capabilities include calculation of slope, aspect, is - lines or contouring, range estimation, perspectives, volumes etc.. Additional functions for determining spatial visibility zones, line of sight are also provided. Network module of ARC / INFO performs two general categories of functions - network analysis and address geocoding. Network analysis for optimal path determination and resource allocation analysis is possible. Geocoding module allows for associating addresses to line networks and determining the spatial framework of a addresses in an application. There are several other modules with advanced concepts of data base, web application and open GIS the details of which are not covered here.

The data models to represent the real world scenario describe various features of a GIS package. The data models, which are in practice, are vector data models, raster data models and TIN (Triangulated irregular network). The GIS software packages like Arc /Info uses the vector data model as shown in Fig 6.6. The details of these data models of ESRI are given below. A vector data model consists of points, lines and polygons to represent any feature on the surface of the earth. The method of representation of vector features in terms of x, y coordinates is given in Fig. 6.7. The vector representation is nearer to real life situation.

Thc topology, which explains the neighborhood of any point, line or polygon, is shown as an example in Fig. 6.8. The raster or grid representation for a typical earth surface is shown in Fig. 6.9. These examples are described in detail in Arc /Info ESRI publications.

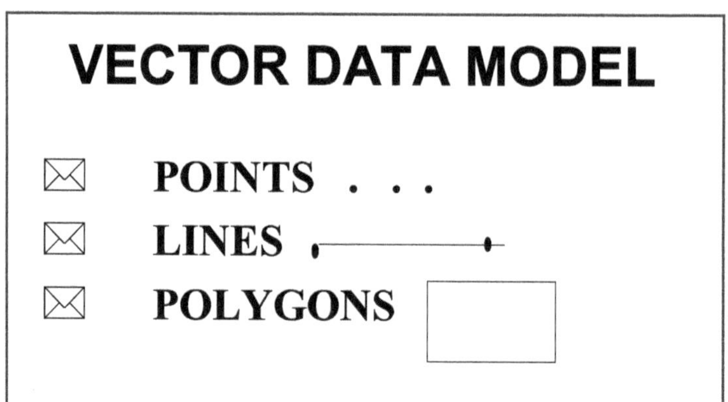

Fig. 6.6 Vector Data Model

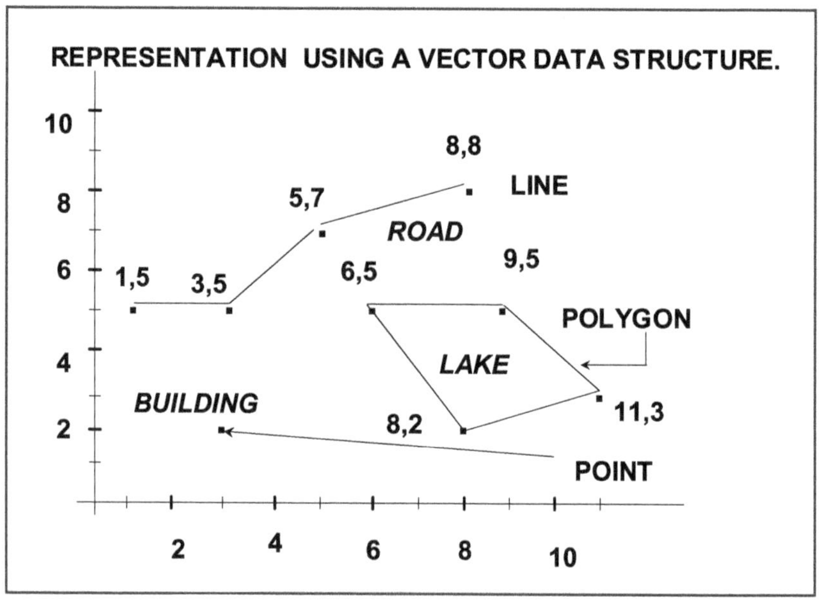

Fig. 6.7 Vector data structure.

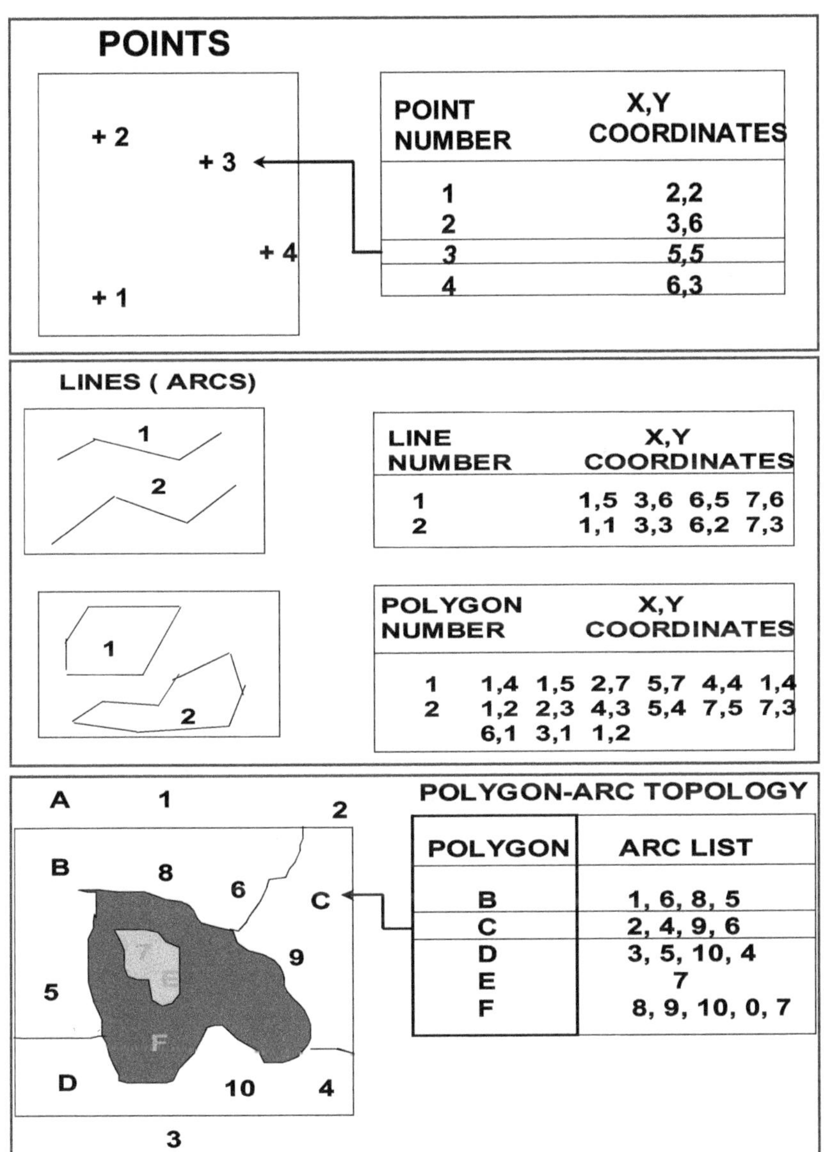

Fig. 6.8 Topology for point, line and polygon.

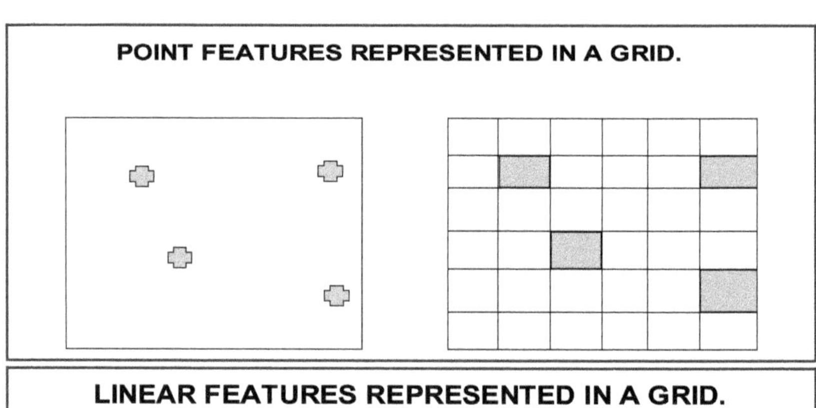

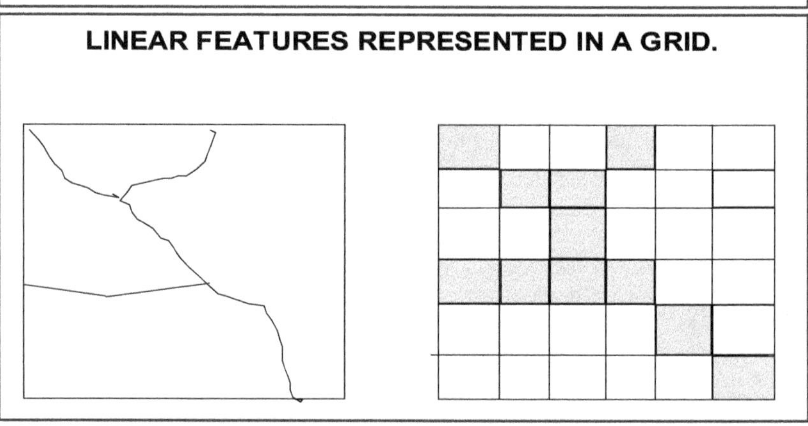

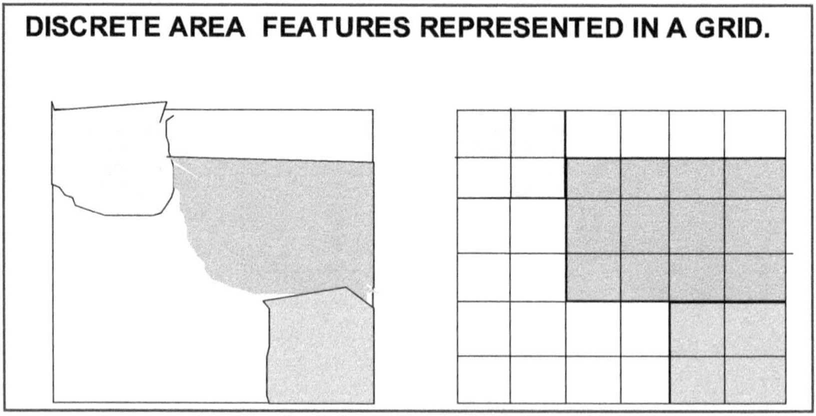

Fig. 6.9 Grid/raster representation of point, line and area features.

Gis for Watershed Management

The watershed has emerged as planning unit for development and management programs. Maintaining the soil factuality is one of the aims of watershed management programs. Therefore amount of soil loss, causes of soil erosion must be found out. Geographic information system is a useful tool for estimating soil erosion effectively. In the present study universal soil loss equation has been incorporated in GIS to calculate soil erosion. The workflow for GIS application based on satellite data and field observations and empirical model development is given schematically in the Fig. 6.10.

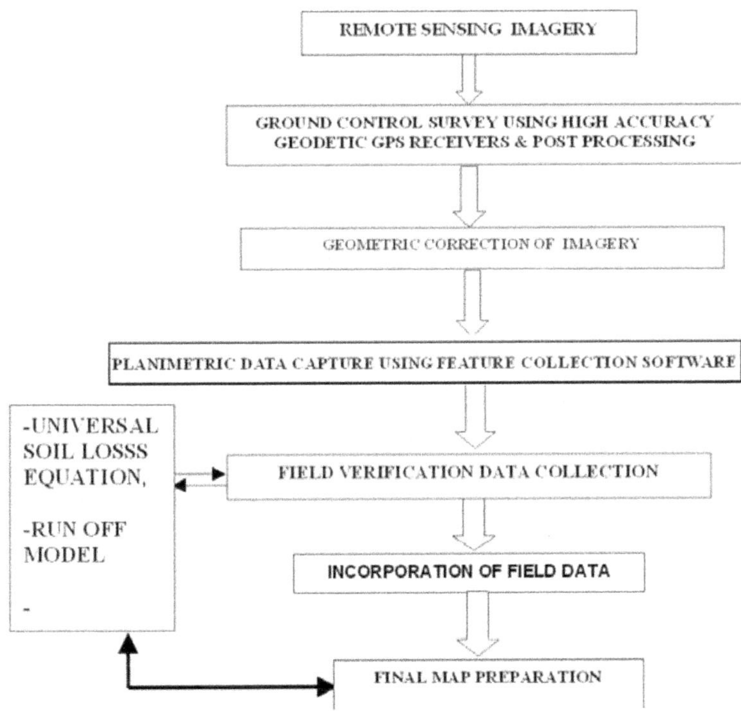

Fig. 6.10 Workflow diagram of the GIS layer creation approach.

These efforts require generation of micro level land use and geomorphic features as layers. The existing data on land and water features are to be obtained from related records and to be converted to digital form to be used in conjunction with thematic maps of resource data. The micro level land use features and the other infrastructure are then verified on the ground and marked on the map to generate a comprehensive information source.

The GIS has unique features to relate the point, linear and area features in terms of the topology as well as connectivity. For example there could be two micro-watershed areas which are adjacent to each other or there could be a feature that requires separate delineation or for identification or exclusion. In this back ground the area- area relationships as given by arc-info GIS software would be of immense use. These sorts of relationships are shown in Figs 6.11 and Fig. 6.12. These figures include all possible combinations like watershed boundaries along with linear features like drainage or road and point features like water outflow / inflow points.

Area-area relationships

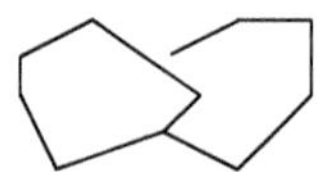

(a) Two watersheds adjacent to each other

(b) A microwatershed feature within a higher order watershed

Area-line relationships

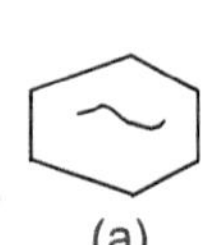

(a)

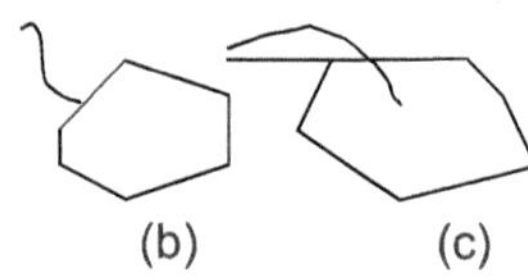

(b) (c)

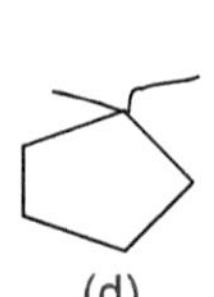

(d)

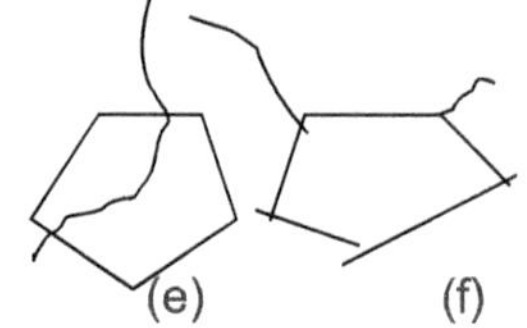

(e) (f)

(a) A Linear feature within a watershed polygon
(b) A Linear feature ends at a watershed polygon
(c) A Linear feature ends in a watershed polygon
(d) A Linear feature touches a watershed polygon
(e) A Linear feature intersects a watershed polygon
(f) A Linear feature acts as a border of a polygon

Fig. 6.11 Typical examples of GIS topology build up for watershed features for area-area and area-line features.

Point-line relationships Point-area relationships

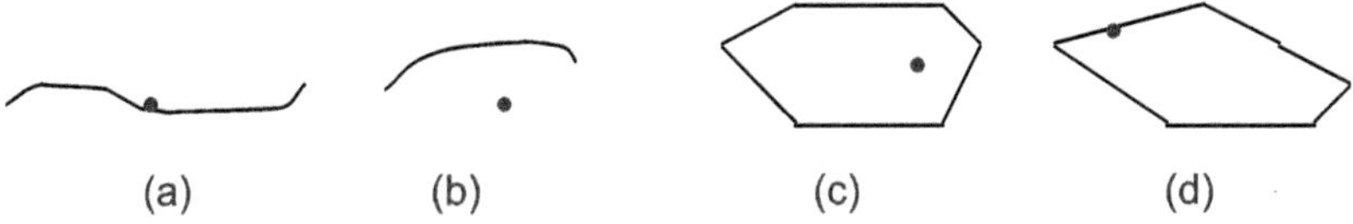

(a) A point feature on a stream like feature

(b) A point feature surrounding /beside a stream like feature

(c) A point feature within a watershed polygon

(d) A point feature on border of a polygon

Fig. 6.12 Typical examples of GIS topology build up for watershed features for *point line* and *point area* features.

Watershed Characteristics and GIS Database Creation

The important watershed characteristics identified for the purpose of defining and managing are as follows:

Size : The small watershed behaves different from the larger ones in terms of the relative importance of various phases of the runoff phenomenon. In small catchments the overland flow phase is pre-dominant over the channel flow. Hence the land use and intensity of rainfall have important role on the peak flood. In large watershed the channel flow phase is more predominant. Further the larger the area, the greater will be the heterogeneity in the soils, vegetation, slope, land use, precipitation characteristics etc. The size of watershed can be determined through GIS. First of all watershed should be delineated by using contour and drainage map.

Shape : Watersheds have many shapes e.g. it could be rectangular, square, palm shape, oval etc. The shape controls the length, width ratio which affects the runoff characteristics. For example, the longer the watershed, the greater is the time of concentration. The longer the time of concentration, the greater the time available for the water to infiltrate, evaporate and got utilized by vegetation. The shape of a watershed may be described by a shape index Sw:

$$Sw = \frac{L}{W} = \frac{L^2}{A}$$

where, Sw = Watershed index,

L = Length of the watershed along the main stream from the watershed outlet to the most distant ridge.

W = Average width of watershed,

A = Area of watershed

Shape of watershed is the output of delineated watershed boundary map by using contour and drainage as input.

Slope : This is a very important characteristics of a watershed. It affects the time of concentration, infiltration opportunity time, and runoff and soil loss. The average watershed slope in percent may be determined from a map by the following formula :

$$S = \frac{MN}{A} \times 100$$

where, M = Total length of contours within the watershed,

N = Contour interval and

A = Watershed area

For very small watersheds the average slope can be taken as the ratio of elevation difference between outlet and most distant ridge to average length of watershed. Slope map can be derived from Digital Elevation Model.

Soil Cover Complex

A combination of a specific soil and a specific vegetative cover is referred to as a soil cover complex and a measure of this complex can be used as a watershed parameter in estimating runoff. The hydrologic properties of a soil or a group of soils are an essential factor in the hydrologic analysis of watershed data. Soils can be classified according to their hydrologic properties if considered independently of watershed slope and cover. Four major soil groups are recognized for the primary classification of watershed soils. These are essentially based on infiltration rates (low, moderate and high), drainage characteristics, soil depth, texture (fine, coarse etc) and rate of water transmission. Hydrologic soil group map can be created by reclassifying soil map. The soil map can be derived by using physiographic association method or by using Remote Sensing in conjunction of ground truth. In physiographic association method, aspect map, land use map and elevation map are crossed and then the composite map is reclassified as soil map.

Vegetation Cover

Vegetation influences water movement behavior significantly it affects the infiltration rate, soil erosion evapo-transpiration, sediment production etc. Remote sensing is a very effective tool to get this information. FCC can be divided into various landuse/ land cover. NDVI gives the vegetation intensity, which can also be used to assess the vegetal status of land.

Drainage

This is another important factor which influences the watershed behaviour. The drainage density is defined as length of drainage channels per unit area. A large drainage density creates situation conducive for quick disposal of runoff down the channels. In watershed where drainage density is small overland flow is predominant.

Climate

Climate parameters like precipitation, humidity, temperature, wind etc. affect the functioning of watershed. Intensity, duration & frequency greatly affect the watershed hydrology. These parameters are the input for hydrological models in GIS environment.

Time of Concentration

This is defined as time which takes for water to travel from the most distant point of a watershed to the watershed outlet. Following formulae is generally used to determine the time of concentration.

$$T_c = \frac{L^{1.15}}{7700H^{0.38}}$$

where, T_c = Time of concentration, hr.,

L = length of the watershed along the mainstream. This can be calculated through distance calculation in GIS environment

H = Difference in elevation between the watershed outlet and the most distant ridge in ft.

H can be calculated through DEM. However, differences in elevation due to over falls, rapids, or other sudden drops should be subtracted from H. Various watershed characteristics can obtained from Remote Sensing toposheets and base map (IIRS – Lecture Notes on Water Resources Management) Accurate prediction of runoff is difficult as it depends upon several factors. There are several methods like rational method, SCS method etc., evolved after field observation for estimating the rate or the maximum rate of runoff that could occur from a particular catchment.

In rational method, the peak rate of runoff is estimated through a model which is dependent on intensity of rainfall (mm/hr) for a duration equal to the time of concentration and for the given frequency, runoff coefficient, area of the catchment in hectares or acres. Runoff coefficient C is defined as the ratio of the peak runoff rate to the rainfall intensity. Values of C for different slopes and land use conditions are determined from field observations. The value of the intensity of rainfall to be used in the empirical model should be calculated for the period equal to the time of concentration of the catchment. When the duration of rainfall equals the time of concentration, all parts of the catchment will be able to contribute to the discharge at the outlet and as such the discharge will be maximum. The peak rate of runoff is to be determined with the help of Geographic Information System. The catchment characteristics may be varying and this can be considered by assigning weightages for different parts of the catchment area.

The method that Soil Conservation Services (SCS) proposed in 1986 is perhaps the most commonly used method all over the world for estimation of runoff resulting from rainfall over small watersheds. To estimate runoff using SCS model following input GIS layers are required

- Landuse map
- Soil series map
- DEM for Terrain information
- Watershed boundary Map,
- Rainfall Map,
- Curve Number (CN) Map

The SCS approach involves the use of simple empirical formulae and readily available tables and curves. The empirical equation requires the rainfall and a watershed coefficient as inputs. The watershed coefficient called the curve number (CN) is an index that represents the combinations of hydrologic and geomorphic and land use categories. The remote sensing derived thematic map / information is used for estimating the land cover distributions, hence these provide useful input support for SCS model. During a given rainfall, water is continually being abstracted to saturate thc upper levels of the soil surface; however, this saturation or infiltration is only one of many continuous abstractions. Rainfall is also intercepted by trees, plants, and roof surfaces, and at the same time is evaporated. Once rain falls and fulfills initial requirements of infiltration, natural depressions collect falling rain to form small puddles, creating depression storage. In addition, numerous pools of water forming detention storage build up on permeable and impermeable surfaces within the watershed. This stored water gathers in small rivulets, which carry the water originating as overland flow into small channels, then into larger channels, and finally as channel flow to the watershed outlet. Remote sensing and GIS play a very important role in watershed management. With the help of remote sensing data, change detection of land use / land cover is possible. The high resolution data which is available in panchromatic mode can be used to increase the accuracy of land use / land cover classification. Digital elevation model provides the perspective view of watershed. Slope and aspect map can be created by taking DEM as input in an GIS environment. The overlaying of drainage map over DEM and land use maps help in identification of sites for location of water harvesting structures. (IIRS 2002)

The most commonly used method for estimation of soil loss is Universal Soil Loss Equation (USLE). The essence of the USLE is to isolate each variable and reduce its effect to a number so that when the numbers are multiplied together the result is the amount of soil loss. The equation takes into account the influence of the total rainfall energy for a specific area rather than rainfall amount. Total rainfall energy can be readily computed for localized areas from existing Weather Bureau data. The USLE predicts the longtime average soil losses from a specified land in a specified cropping and management system. The equation predicts only the losses from sheet and rill erosion under specified conditions. With appropriate selection of numerical values for various soil erosion variables, the equation will compute the average soil loss for a cropping system, for a particular crop year in a rotation, or a particular crop stage period within a crop year. It computes the soil loss for a given site as a product of six major factors whose most likely values at a particular location can be expressed numerically.

The USLE can be presented as :

$$A = R * K * LS * S * C * P$$

Where A is the average annual soil loss in tons /acres.

R is the rainfall factor, which is a measure of the erosion force of the rain

K is a soil erodibility factor; the erosion rate per unit of erosion index for a specific soil

LS is slope length factor, the ratio of soil loss from the field slope length to that from a 22.1 meters length on the same soil type

S is slope gradient factor, the ratio of soil loss from the field gradient to that from a 9 percent slope on the same soil type and slope length

C is a cropping and management factor,

P is the supporting conservation practice, such as terracing, strip cropping, and contouring

The rainfall erosion potential. -R-factor can be determined by using following formula,

$$R = -8.12 + 0.562 * \text{Annual Rainfall}$$

Rain gauges installed at various meteorological observatories give depth of rainfall at that place. This point information can be converted to spatial distribution by Theisean polygon method in GIS environment. Once this Theisean polygon map is derived, then by above formula, R-map can be drawn.

The soil – erodibility factor K defines the inherent erosion potential of the soil. It is expressed as the soil loss in tons per acre for each unit of rainfall erosion index for the locality and for continuous fallow tillage on a 9 per cent slope, 22 meters (73 ft) in length. Standard K values have been established for only a few soil types in the country where actual soil loss measurements have been made. Research now in progress may eventually provide criteria for estimating readability. Numerous factors influence erosion of cohesive soils, including but not limited to, texture, grain-size distribution, nature of clay minerals, thickness and permeability of strata and organic content present estimates of soil readability must be made on the basis of know erosion characteristics of the soil.

$K = M^{1.4} (10^{-4}) (12 - a) +.25 (b - 2) + 2.5 (c - 3)$ where

M = particles size parameter

a = percent organic matter,

b = the soil structural code used in soil classification and

c = profile permeability class

The slope length steepness factor or soil loss ratio, LS is determined by dividing the existing length and steepness by the standard 9 per cent slope, 22 meters (73 ft) in length. The soil loss ratio for many given slope conditions can be readily determined by means of the following revised equation :

$LS = (L/22)m\ (10.8 \sin\ \ 0.03)$ for slopes $< 9\%$

$LS = (L/22)m\ (16.8 \sin\ - 0.5)$ for slopes $> = 9\%$

L = Length of slope in meters

The cropping management factor, C is the expected ratio of soil loss from land cropped under specified conditions of soil loss from clean tilled fallow on identical soil and slope and under the same rainfall. This item reflects the combined effect of crop sequence, productivity level, length of growing season, tillage practices, residue management and the expected time distribution of erosive rain storm with respect to seeding and harvesting data in the locality. This factor is the most complicated because there is an almost infinite number of different ways of managing the growing crops. In the early system, such as the slope practice equation, a single value for this crop factor was used to give the average effect over the whole season.

The product of RKLS in the Soil Loss Equation computes the soil loss that would occur under fallow conditions. Actual loss from the cropped field is usually much less than this amount. The reduction depends on the particular combination of land cover, crop sequence and management practices. It also depends upon the particular stage of growth and development of the vegetal cover at the time of rain. C adjusts the soil loss estimate to suit these conditions Based on the data from runoff plots of various sizes established in different agro climatic area of the country, a preliminary evaluation of the factor, C has been made. Although a number of crops have been studied but the detail, data of soil loss for various crop stages is rarely available in the published record. In absence of crop stage growth period data, average value of C of the crop has been determined in a number of cases based on the total seasonal soil loss data..

Watershed Characterization of Units Under Veligonda Irrigation Project - A Case History

The upland areas of Prakasam, Cuddapah and Nellore districts in Andhra Pradesh, India have been identified as chronically drought affected areas. Government of Andhra Pradesh has conceived and planned a project to provide relief against drought and famine in the semi-arid, famine effected areas of these three districts by supplying 43.5 TMC of floodwaters, from river Krishna, which are expected to be the surplus flows of the Srisailam Reservoir. This project is called Veligonda Irrigation Project. The project can be constructed as a network of canals, tunnels and reservoirs, feeding each other and the command area. The Veligonda Project proposes to provide irrigation facility to 1.772 lakh ha. (4.38 lakh acres) and drinking water to about 15 lakh people in the districts of Prakasam, Cuddapah and Nellore (one lakh = 100,000). The estimated Gross command area under the project is 4.06 lakh ha. and the cultivable command area is 2.40 lakh ha. The ayacut proposed is 1.772 lakh ha. The ayacut is spread over 29 mandals (county) in the three districts. The construction of reservoirs may lead to submergence of settlements, agricultural lands and forest areas and it is important to ascertain the ecological impact. The life span of the reservoir depends on efficient management of the vegetation cover and soil cover in the catchment. In this background it is necessary to make a precise estimation of these effects and planning for compensatory afforestation and socio-economic rehabilitation. The Remote Sensing technology due to its synoptic, repetitive and multispectral coverage facilitates mapping of natural resources information on vegetation, land cover, land use, geology, hydrogeomorphology and soils. The specific objectives of the study are as follows:

- To Map surface water bodies and drainage on 1:50,000 scale for the catchment areas
- To Map land use / land cover, Hydro geomorphology, and slope on 1:50,000 scale for the catchment areas.
- To estimate sediment yield and submersible areas under reservoirs.
- To prepare catchment area treatment plans on 1:50,000 scale.
- To map land use / land cover , Ground water potential and Soils maps on 1: 250,000 scale for the command area.
- To integrate these thematic maps as layers under a GIS environment

The following data were used for thematic mapping on 1: 50,000 and 1: 250,000 scales:

- Geocoded Imagery of .Indian Remote Sensing Satellite (IRS- 1B/ 1C- LISS II / III for February/March 1999 and October 1992.
- Survey of India topographical maps on 1:50,000 scale and on 1:250,000 scale.

The streams drained into the reservoirs have been demarcated based on Survey of India toposheets on 1:50,000 scale. The base maps for the study area are prepared showing major roads, settlements, rivers/tanks and forest boundaries on 1:50,000 scale. The base maps are superimposed on IRS-1B LISS- geocoded (FCC) products at 1:50,000 scale and various land use/land cover categories viz., deciduous forest, open forest, degraded / scrub forest, Wastelands such as land with scrub, land without scrub and irrigated, un-irrigated double crop and water bodies were delineated and mapped using visual interpretation technique. The areas of various categories were calculated and statistics are presented. The submergible areas under each reservoir were mapped and unit wise land use/ land cover areas are estimated. The same procedure is adopted for command area mapping on 1: 250,000 scale. The hydro geomorphology mapping was made on both 1: 50,000 and 1:250,000 scales. The land use / land cover analysis for the 5 reservoir catchments was carried out with a view to assess baseline status of land use/land cover of the catchments and submersible areas to account for the ecological conditions. These maps have been used to study the sediment yield in the catchments and to prepare catchment area treatment plans. The prime aim of catchment area treatment plans preparation is to suggest suitable treatment measures, which will reduce the rate of erosion in the directly drained catchment area and to retard the rate of siltation. The plan also includes areas identified for afforestation of the under stocked areas and increasing the density of existing vegetation cover. This would help maintenance of environmental balance, preservation of bio-diversity and maintenance of the gene pool in the existing area and to bring about eco-development of the area.

The land use / land cover, slope, soil and drainage maps along with rainfall data were considered for the preparation of treatment plan on 1:50,000 scale. The area of each unit, which is proposed for the treatment, has been estimated and presented. The treatment plan comprises essentially of two components such as biotic treatment with soil and moisture conservation measures and engineering and gully control treatment works. The areas identified for the biotic treatment with soil conservation measures primarily fall under the land without scrub, land with scrub, open forest, degraded/scrub forest and kharif un-irrigated lands. Any biotic treatment for land with scrub and land without scrub is mainly to conserve soil and water, and minimize soil losses from erosion with plants, shrubs and legumes/grasses and afforestation in the gap areas with social forestry or horticultural crops. All the wastelands that do not qualify for placement under any of the wasteland categories outside the forest area are included under this class. Besides scrub lands, other lands with soils that are too shallow, skeletal, or otherwise chemically degraded, lands with extremes of slopes, severely eroded lands and lands subject to excessive aridity, etc., are placed in this group viz. Biotic treatment for land with and without scrub. Biotic treatment for Deciduous open and scrub forest /degraded forest is for rejuvenation and regeneration of degraded forests and afforestation in the gap areas for eco-restoration. To retard siltation from directly drained areas into the reservoir, the lands situated below 0.6 m (2 feet) and above 0.6 m of the FRL have been delineated from satellite imageries and recommended for green belt plantation

The gully control works/ engineering treatment measurements are recommended in the treatment areas mainly to control the sediment from the catchments as well as to increase the ground water recharge. Engineering measures include the construction of hydraulic structures like check dams and water harvesting ponds, gully plugs, bank protection etc., Five types of such works viz., brush wood dam, gully plugs, rock fill dams, check dams, minor tanks and stream bank stabilization are proposed in catchments based on the details inferred from satellite imageries and field observation. The soil map prepared using satellite imageries for the project area on 1:250,000 scales provide information on soils of the area. The study provides baseline data and action plan that would help the field officer to implement schemes for reduction of rate of siltation of reservoir under appropriate management strategies. The study also demonstrates the utility of remote sensing technology for command area development and catchment area treatment. The remote sensing data can also be used to derive engineering parameters. As an example, calculation of the vegetative cover is given which will be required for estimation of some of the parameters like vegetative cover to estimate the sediment yield of the catchments The details of Land use / land cover derived from remote sensing data are given below as examples of a typical case of application:

Kharif unirrigated	:	123.975 sq. km.
Kharif irrigated	:	1.150 sq. km.
Deciduous forest (dense)	:	176.000 sq. km.
Deciduous forest (open)	:	15.225 sq. km.
Degraded/Scrub forest	:	29.850 sq. km
Land with scrub	:	22.425 sq. km

The above parameters are regrouped to calculate the *vegetative cover factor(Fc)*

Fc = Vegetative cover = 0.2 F1 + 0.2 F2 + 0.6 F3 + 0.8 F4 + F5 .

The values of F1 to F5 are calculated from the corresponding thematic maps.

F1 = Area under reserve and protected forest

= Deciduous forest (dense +open) = 176 + 15.225

= 191.225

F2 = Unclassified forest area = 0 (zero)

F3 = Cultivated area = Kharif unirrigated + Kharif irrigated

= 123.975+1.15+0.25 = 125.375

F4 = Grass and pasture land = 0 (zero)

F5 = Wasteland = Land with scrub + land without scrub + Scrub forest

= 29.85 + 22.425 + 5.9 = 58.175

The digital thematic layers of land use help calculation of parameters F1 to F5. Thus the extent of vegetative cover can be created by integration of the values of F1 to F5. The performance of an irrigation project is judged on the basis of actual utilization of irrigation potential created and increase in agricultural production. The area commanded by the irrigation canal network has to be scientifically developed to receive the water. Basically the prime task for field officers and engineers concerned with the command area development, envisages land grading and shaping. The land surface to be irrigated is expected to provide for controlled and safe flow over it. The land is to be graded to the required longitudinal and cross slopes. The land grading for irrigation should meet both grade criteria and the grading criteria. The former aims at efficient application of water in the field such that the deep percolation (beyond the root zone) and the surface run off are minimized and uniform wetting of the soil profile is attained. The grading criteria on the other hand concerns with the minimum earthwork, appropriate cut fill ratio and limited cuts so as not to expose the sub soil. The land grading is a costly affair. In order to minimize the cost, existing topography should be exploited to the maximum extent. The basic approaches in land grading and field layout on gently sloping lands, with the fields laid perpendicular to the contour lines, to provide the necessary down field gradient for irrigation. On strongly sloping lands the benches are constructed across the slopes on which the graded fields are laid out. The farm irrigation system in the outlet commands should consist of field channels, lined or unlined, along with control structures to convey the water to the fields. A farm irrigation system should have counterpart farm drainage system for removal of excess water to prevent water logging and salinity. The farm drainage system can be either a surface drainage system comprising open drains or sub surface drainage system comprising tile drains or open drains. The surface drainage removes the excess rain or irrigation water from land surface, while the later keeps the water table below the root zone. Farm irrigation structures are used for water control from the outlets. They are needed to divide flow of water (division structures), control velocity (drop structures) deliver water without erosion (check structures) measure water (weirs, flumes) and deliver correct amount of water (turnouts, siphons, valves and gated pipes) Inadequate distribution of water can be corrected through the use of irrigation structures and improve water use efficiency. The benefits derived from properly installed irrigation structures are in the form of reduced erosion on sloping lands, control of amount of water being delivered to fields, reduce water course deterioration and increase equitable water distribution among the farmers. It helps control of waste and seepage water and reduction in water logging and salinity. The area commanded in the present study is gently sloping. The on-farm development warrants systematic land development and construction of bunds, provision of surface drains, improvement of existing drains, construction of irrigation structures, training of farmers in irrigated agriculture, provision of credit facilities through banks to enable the farmers to purchase inputs

GIS Issues

A brief account of issues related to irrigation and command area development are narrated here primarily to bring to attention those measures that can be handled by remote sensing data and integration with collateral data so as to develop a GIS for command / catchment area development. The basic issue for any application related to GIS essentially is concerned with input data. The input data is both in spatial as well as non spatial form. Maintenance and analysis of spatial data, maintenance and analysis of non-spatial data, integration of spatial and non spatial data are the three major components of any effort related to development of GIS for watershed management.

In an article of this type particularly while dealing diverse topics like watershed management and GIS several omissions and commissions can be expected because of the vastness of the technology to be explained as well as domain knowledge expected to be the background of the individual trying to understand the issues related to application of GIS.

The issues related to GIS can be restated and or summarized as follows:

- Geographic Information System (GIS), computer system that records, stores, and analyzes information about the features that make up the earth's surface.
- A GIS can generate two- or three-dimensional images of an area, showing such natural features as hills and rivers with artificial features such as canals and existing water harvesting structures, if any.
- Many GIS databases consist of sets of information called *layers.* Each layer represents a particular type of geographic data. For example, one layer may include information on the water bodies in an area. Another layer may contain information on the soil in that area, while another records elevation.
- The GIS can combine these layers into one image, showing how the canals, soil, and elevation relate to one another. Field officers might use this image to determine whether a particular part of a watershed is more likely to crumble. A GIS database can include as many as 100 layers.
- A GIS is designed to accept geographic data from a variety of sources, including maps, satellite photographs, and printed text and statistics. GIS sensors can scan some of this data directly - for example, a computer operator may feed a map or photograph into the scanner, and the computer "reads" the information it contains.
- The GIS converts all geographical data into a digital code, which it arranges in its database. Operators program the GIS to process the information and produce the images or information they need.

References

Burrough, P A (2002) *Principles of Geographic Information Systems for Land Resources*,

Environmental Systems Research Institute, Redlands, Calif., USA (ESRI) (2002) ARC/ INFO/ ARC / GIS Manual

Gregory K. J, & Walling D. E (1973) *Drainage Basin Form and Process: geomorphological approach*, Edward Arnold., U. K.

Perumal A, (1996) *Watershed characterization*. Integrated Mission for Sustainable Development (IMSD) , Rept. of the National Remote Sensing Agency, Hyderabad, India

Pimentel, D. (Ed.) (1993) *World Soil Erosion and Conservation* , Cambridge University Press, U.K.

7

Water Resources Systems Engineering

M. Yacoob

Consultant, Irrigation Management and Training, Hyderabad, India.
(e-mail : myom@rediffmail.com)

Key words : Water resources as a system, linear programming, Allocation, Competing users, Water distribution networks, Decision support system

Introduction

Water resources have an undeniably important role in the economic development of a country. Although, the planning, design and management of water resources involve huge economic costs attendant with a variety of ecological, environmental and social impacts, the most important and quantifiable criterion for decision making on water resources development is the benefits that accrue from it against the costs of its implementation.

Most of the proposed regional water resources developments involve investment in land, reservoirs, pipelines/canals and/or pumping/ power generation facilities and the like. As a result of these capital investments, economic returns may be obtained for long decades. But, both in an economic and ecological sense, these investments are often irreversible; and, therefore, it does not make good sense if all the aspects involved are not thoroughly examined before taking appropriate decisions for helping the development process in line with the set goals of the state. This is indeed a complex job.

However, over the last few decades, certain improved analytical tools and methodologies have been developed to define and evaluate the various alternative approaches which enable to make the best possible decisions. Whether or not they can ensure optimal plans for development and management, they can certainly help define and evaluate – in the face of competing and changing objectives and priorities of different interest groups – in a rather detailed manner, numerous alternatives that represent various possible compromises among conflicting groups, values and management objectives to help identify the possible trade-offs between quantifiable objectives so that further debate and analysis is more informed.

Water Resources as a System

After the introduction of the Linear Programming (Dantzig, 1963) and Dynamic Programming (Bellman, 1957), some landmark break-throughs have been made in the design and analysis of water resources problems for their optimal use. Mass et al (1962), presented their classic work on the 'Design of Water Resources System' while Major & Lenton (1969), dealt with 'Applied Water Resources Systems Planning' and Hufschmidt & Fiering (1966) introduced the 'Simulation Techniques for the Design of Water Resources Systems'.

Other noteworthy contributions on the application of systems analysis approach to water resources problems are the classics of Hall and Dracup (1970), Major (1977), Haimes et al (1975), Haimes (1977), and Loucks et al (1981).

With further advances in the areas of hydrology and stochastic hydrology, water resources economics and applied mathematics, the techniques of systems analysis got further refined for a meaningful analysis of water resources planning, design and operational problems, involving single/ multiple objectives or a hierarchical decision process.

The developments in the computational sciences have further improved the decision-making processes for off-line and on-line/real-time management. For more details, readers are advised to consult the literature on the latest developments.

Systems Approach

The art of systems analysis is

- To identify those issues and concerns which are important and significant; and
- To structure the analysis to throw light on these issues.

The important tool employed - often called as mathematical modeling or optimization or systems analysis - and conceptually depicted in Fig. 7.1, is characterised by :

- A rule which determines whether any particular object is to be considered as a part of the system or of the environment
- A statement of the input and output interactions with the environment
- A statement of the inter-relationships between the elements of the system, the inputs and the outputs, including any external interaction between output and input (feed back).

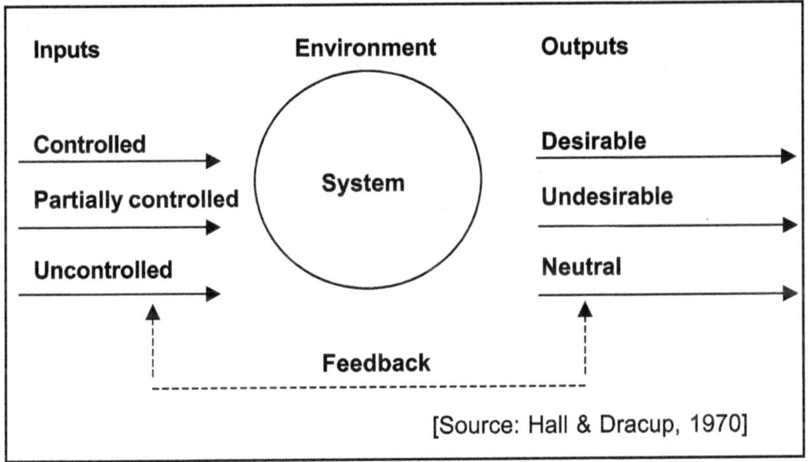

Fig. 7.1 The concept of a system.

In such a characterization of the system, the task of analysis is to modify the controllable/partially controllable inputs so as to maximize the desirable outputs and minimize the undesirable outputs.

Such an analysis is capable of :

- Representing, in a fairly structured and ordered manner, actions among the various control structures and users of a water resources system
- Evaluating the economic and physical consequences of
  (a) Alternative engineering structures,
  (b) Various operating and allocation policies in the face of different assumptions made regarding water resources available in future, technology, costs and social/legal requirements.

What it can do and Cannot do

Although, this methodology cannot define the best objectives and assumptions, it can identify good decisions under the given objectives and assumptions. As such, it is not important to expect the precise results of the system study to have acceptance for implementation; a measure of its success can be seen in what it has been able to do; viz:

1. A beneficial impact on the planning and decision-making process
2. Cause a debate over the proper choices of alternatives in a more informed manner
3. Introduce competing alternative proposals which otherwise would not have been considered.

Requirements

The exact approach to modeling, however, depends upon :

1. The objectives of analysis
2. The data requirement for the evaluation of the project alternatives
3. The time/data/money/computational facilities available for the analysis
4. The skills of the modeler

Thus, the tool of systems analysis has made the decision-making process more informed and objective; and, for this reason, it has gained great acceptability among the planners/decision-makers and the management specialists all over the world.

Optimization Methods in Water Management

Definitions

The systems analyst generally uses certain terms to represent the characteristics of a system. These are briefly described below :

Decision Variables : The partially controllable and controllable inputs.

Policy : The set of decisions that results when each decision variable is assigned a certain value.

Feasible Policy : A policy which does not violate any constraint.

Policy Space : The subset that consists of all feasible policies

State Variables : The variables that define the condition of the system at any given time/space.

System Parameters : The state variables that are varied in a stipulated manner to study their impact on the solution.

Objective : The criterion whereby the effect of any feasible policy on the desirable/undesirable outputs can be gauged.

The objectives may be quantifiable or non-quantifiable.

Model Formulation

In the model for systems analysis, the criterion or the objective function has a very important role to play. The objective function, as has been explained above, is a statement by which the consequences or outputs of the system can be determined, given the policy, the initial values of the state variables and the system parameters. However, such a determination is possible only if the quantitative objectives are fully commensurate. It is, thus, a simplified mathematical statement of objectives, fully commensurate. It should be noted that this objective function is not synonymous with the objectives of the decision-makers even in the simplest of situations.

For example, the Government agencies often indicate that their goal is "the greatest good to the greatest number", which by no means can be described by an objective function unless there is a fixed rule for equating units of 'good' to 'number'.

The problems of water resources development/management are, therefore, not amenable to an exhaustive analysis because of :

1. Non-commensurate nature of policy statements
2. The quantitative functional relationships encountered being incompatible with the general necessary and sufficient conditions for the standard formulations
3. A large number of decision variables involved.

With the above reservations, if we assume that the objective function is a scalar function of

(i) The decision variables Xi, i = 1,2....I

(ii) The state variables Sj, j = 1,2.....J

(iii) The system parameters Pk, k = 1,2....K

If we also assume that the constraints acting on the system can be adequately expressed by a set of M functions g_m (Xi,Sj,Pk), m = 1,2,...M

Then the fundamental problem of systems analysis is to select that particular policy (i.e the particular value of the Xi) which maximizes/ minimizes the objective function subject to the constraints stipulated above. This can be symbolically expressed as

$$V = \text{Max (Min)} [V (Xi,Sj,Pk)] \quad ......(7.1)$$

Subject to the constraints of the form

$$g_m (Xi, Sj, Pk) < 0$$

and $$Xi > 0$$

The eqn. (7.1) is little more than an organization of the ideas. It tells us nothing about how the optimal policy is to be obtained. Further, it usually omits the non-quantitative aspects of the problem, and non-commensurate objectives, which may be reflected by the constraints. The task now, therefore, is to solve the above problem with a minimum of computational difficulty.

Solution Techniques

The solution techniques that can be adopted depend on the nature of the objective function as to whether

(a) It is linear/nonlinear, continuous/discrete, differentiable/non-differentiable

(b) With or without constraints

Some of the popular techniques of analysis are as follows:

Calculus Method : Applicable when the functions are continuous and continuously differentiable with or without constraints

Linear Programming : When the objective function and the constraining conditions are linear or piece-wise liberalization is possible

Dynamic Programming : (i) For functions linear or non-linear, continuous or non- continuous, differentiable or otherwise (ii) Sequential problems

Non-linear Programming : For non-linear functions

Besides these, several other programming techniques, such as integer programming, goal programming, multi-objective programming, etc, have also been developed to solve certain complex type of water resources problems.

Applications

As has been stated elsewhere, systems analysis is most useful in planning problems, besides its applications for management of resources in an optimal manner.

In the present discussion, the linear programming formulation and its application to problems of allocation of water among different competing users – a problem often encountered in the planning phase – is described.

Allocation of Water to Competing Users - Linear Programming Approach:

L.P. Formulation

As has been stated earlier, when the objective function and the constraint equations are linear functions, a linear programming techniques can be adopted to solve the problem. In that case, the earlier formulation can be re-written as

$$\text{Max } V = V(x_1,\ldots x_n) = c_1x_1+c_2x_2+\ldots\ldots c_n x_n$$

Subject to

$$g_1 (x_1,\ldots\ldots x_n) = a_{11} x_1+a_{12} x_2+\ldots\ldots+a_{1n}x_n \leq b_1$$

$$g_2 (x_1,\ldots\ldots x_n) = a_{21} x_1+a_{22} x_2+\ldots\ldots+a_{2n}x_n \leq b_2$$

$$\cdot g_m (x_1,\ldots\ldots x_n) = a_{m1} x_1+a_{m2} x_2+\ldots\ldots+ a_{mn} x_n \leq b_m$$

This can then be solved by the linear programming techniques.

This technique is eminently suitable for solving the water allocation problems to competing users. This is explained through two examples as described below :

Two Sources, Three Users Allocation Model

Let the sources be :

1. Surface water from a nearby reservoir or stream supplied through canal network/ pipeline
2. Pumped water from groundwater basin.

Let the users be :

1. Municipal & local industrial demand
2. Agricultural demand
3. Recharge of groundwater basin.

This system represents an urban community overlying a groundwater basin. Therefore, the physical constraints and economic characteristics of both the groundwater reservoir and the overlying surface-water facilities influence the optimum management of the system.

Let the unit costs of water under various conditions be defined as follows :

c_s = unit cost of surface water delivered for municipal and industrial use / agriculture

c_t = unit cost of treatment to domestic standards

c_b = unit cost of boosting to required supply-line water pressure

c_m = unit cost of short-term regulatory storage for municipal /industrial supply

c_r = unit cost of artificial recharge

c_l = unit cost of local supplies

c_p = unit cost of pumping ground water

Let the sources be designated by index i, i = 1, 2

Let the destinations be designated by index j, j = 1, 2, 3

Then the overall unit cost-coefficients c_{ij}, i = 1, 2; j = 1, 2, 3 will be as follows :

$$c_{11} = c_s + c_t + c_b + c_m + 0 + 0 + c_l + 0$$

$$c_{12} = c_s + 0 + 0 + 0 + 0 + 0 + 0 + 0$$

$$c_{13} = c_s + 0 + 0 + 0 + 0 + c_r + 0 + 0$$

$$c_{21} = 0 + c_t + c_b + c_m + 0 + 0 + c_l + c_p$$

$$c_{22} = 0 + 0 + 0 + 0 + 0 + 0 + 0 + c_p$$

$$c_{23} = 0 + 0 + 0 + 0 + 0 + 0 + 0 + 0$$

Let the maximum quantities available from the two sources be represented by X_1 and X_2 respectively.

Let the minimum demands to be met for each user be represented by D_1, D_2 and D_3 respectively.

Let the ground water storage be maintained between a minimum of 0 and a maximum of $G_{max.}$

Let the maximum quantity that can be pumped from ground water in any time period be less than $\Delta_{i.}$

With the above assumptions a linear programming model for the allocation of resources could be formulated in the following manner :

$$\text{Min } Z = S^{T}{}_{i=1}\, S^{M}{}_{j=1}\, S^{N}{}_{k=1}\, c_{jk}\, x_{ijk}$$

Where i refers to the time period, $i = 1,2, \ldots\ldots T$

j refers to the sources, $j = 1, \ldots\ldots\ldots M\ (= 2)$

k refers to the users, $k = 1, 2, \ldots\ldots N\ (= 3)$

Subject to

(a) Limitations on source of supply

$$\Sigma^{N}{}_{k=1}\, x_{ilk} \leq X_{i1}$$

$$\Sigma^{N}{}_{k=1}\, x_{i2k} \leq X_{i2}$$

(b) Satisfaction of demands

$$\Sigma^{M}{}_{j=1}\, x_{ij1}{}^{3}\, D_{i1}$$

$$\Sigma^{M}{}_{j=1}\, x_{ij2}{}^{3}\, D_{i2}$$

$$\Sigma^{M}{}_{j=1}\, x_{ij3}{}^{3}\, D_{i3}$$

(c) Ground water level

$$G_i \leq G_{max}$$

(d) Ground water drawn in any time period

$$\Sigma^{N}_{k=1} \quad x_{i-1\,,\,2,\,k} - \Sigma^{N}_{k=1}\, x_{I\,,\,2,\,k}\, \Sigma\, \Delta_i$$

The solutions of this formulation will give the best allocation feasible.

Allocation of Crop Areas in a Project Command

Let

i = 1,2, ...n (n : no of feasible crops)

j = 1,2, ...m (m : no of time periods)

A = Total area to be irrigated under the project

D = Total water available for use

B_j = Capacity supply in time period j

X_{ij} = Crop water requirement of i th crop in j th time period

p_i = Net value of produce per unit area of i th crop

y_i = Irrigated area under i th crop.

Then the objective function of the problem can be written as:

$$\text{Max } Z = \Sigma^{n}_{i=1}\, p_i\, y_i$$

Subject to the constraints of

(a) Supply capacity

$$\Sigma^{n}_{i=1}\, x_{ij}\, y_i \leq B_j \qquad j = 1,2, \ldots\ldots.m$$

(b) Area available for irrigation

$$\Sigma^{n}_{i=1}\, y_i < A$$

(c) Total demand

$$\Sigma^{m}_{j=1}\, B_j < D$$

(d) Crop areas to be non- negative

$$y_i > 0 \qquad i = 1,2, \ldots\ldots n$$

(e) Period-wise demand on the system

$$\Sigma^{n}_{i=1}\, x_{ij}{}^{3}\, y_j \qquad j = 1,2, \ldots\ldots m$$

(f) Restrictions on crop areas if any

$$A_i^{min} < y_i < A_i^{max} \qquad i = 1,2, \ldots\ldots n$$

A solution of the above formulation will give the optimal areas under each crop under the project (Optimal Cropping Pattern). This is demonstrated through a case study (Ramulu & Yaqoob, 1987) that follows.

Sriramsagar Project – A case study

Sriramsagar Project is a major irrigation project constructed across Godavari river to bring the parched lands of Telangana region in Andhra Pradesh under irrigation. In the ultimate stage the total area covered by the project comes to 15.022 lakh acres. It is expected to have a net utilizable yield of 149.24 TMC (after meeting the reservoir losses of about 22.47 TMC)

Cropping Pattern Adopted

The cropping pattern for the Project has been under examination based on soils, climate, food habits and local practices, etc. However, the decision about the choice of the best cropping pattern under the prevailing conditions does not appear to have been based on any scientific method of analysis. This necessitated the present study to evolve a suitable cropping pattern so that the Project performs optimally.

Optimal Cropping Pattern

The cropping pattern could generally be planned either for best return to individual farmers (micro-level) or to conform to goals set forth by regional/ national agencies in respect of food production and other needs without regard to the benefits accruing to individual farmers (macro-level). In the study, the macro-level approach has been adopted in evolving an optimal cropping pattern.

In keeping with the above, in addition to the approved cropping pattern the following alternatives have also been considered to locate the best solution for the Project.

Alt. – I An unrestricted pattern of cropping fully utilizing the available water and the capacity of the canal system

Alt. – II Crop areas to be not less than those developed in 1st phase of the project fully utilizing the available yield and the carrying capacity of the canal system

Alt – III Crop pattern to make the region self-sufficient in food grain production fully utilizing the available yield and the carrying capacity of the canal system.

Project Performance

The optimal cropping pattern emerging from the study under all the alternatives has been considered. It is seen that the Project alternatives considered violate one constraint or the other making it necessary to review the whole problem for selection of the best alternative. However, under identical conditions, Alt. III considered is seen to be a better alternative than the cropping pattern adopted for the Project, deserving further careful consideration.

Conclusions

The broad conclusions of the study are summarized below :

1. All the alternatives considered utilize the water and land resources available to the full. Therefore, all the alternatives considered prove to be satisfactory on that count.

2. The imposition of constraint on minimum area to be developed under each crop proposed (Alt. II) is seen to reduce the benefits by about 5%. Such a constraint may be necessary to take care of the areas already developed and its impact is quite insignificant.

3. The constraint on production of food grains to make the region self-sufficient (Alt. III) reduces the benefits by about 30%. It is, therefore, necessary to include the cost of importation of food grains and cost of export of commodities also to determine the true impact on the performance of the project.

4. For the Project to perform optimally, it is desirable to develop an optimal operation policy consistent with the storage capacity of the reservoir to ensure supplies as required under the optimal cropping pattern for the Project.

   As the yield of water on year to year basis is likely to change, it would be necessary to predict the yield and evolve an optimal cropping pattern and the corresponding optimal operation policy adopted to ensure the best utilization of the resources on an year to year basis.

5. The study could be extended to the entire system of Sriramsagar – Kadam – Lower Manair projects for a better development and utilization of water and land resources available under these projects collectively in the basin.

6. For integrated development of land and water resources (both surface and sub-surface) in an optimal manner, the groundwater potential could also be determined and exploited so that the associated evils of irrigation, like water-logging and salinity, are eliminated and the benefits of irrigation maximized.

Design of Water Allocation and Water Distribution Networks

General

The planning and design of water resources system depends on the availability of the water resource. The sources could be many, namely :

(i) Precipitation water which flows through streams, rivers, etc

(ii) Water stored in the reservoirs/water bodies (natural/man-made), etc

(iii) Soil water i.e the water trapped in the upper layer of the soil as soil moisture.

Therefore, the question of how much water and from where it must be drawn, always assumes great significance in a judicious planning/design/ management process for its optimal use without causing adverse ecological/ environmental problems.

User's Demands & Allocation

The primary water use subsystems comprise, agriculture, municipal and industrial use. The water allocation problems can be viewed in two ways:

(a) How much water should be drawn from each source

(b) How much water should be allocated to each water-use subsystem such that

- Healthy ecological/environmental conditions are ensured
- The benefits accruing to the development goals of the nation (various sections of the people) are maximum at minimum investments.

Having decided as to how much to draw from where and how much to supply to each user subsystem, it requires to develop a storage/conveyance plan to enable distribution of water to various uses most effectively to achieve the objectives set forth.

Distribution System

The process of distribution of water is neither a one-time affair nor a point-process. The demands of the users concerned have to be met in space (i.e from the farmer's field to the project command) as well as in time (i.e over days/weeks/months/seasons, etc).

Keeping the above in view the whole process may be conceived in the manner depicted in Fig. 7.2.

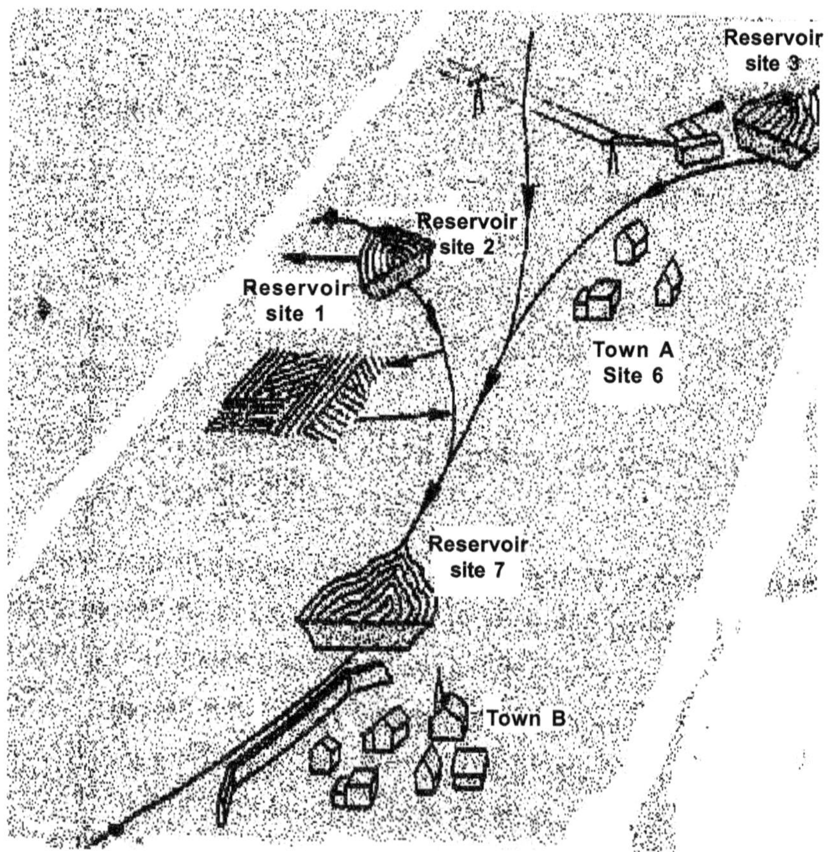

Fig. 7.2 Diagram showing the sources and users.

The Water resources systems may therefore comprise :

(i) ***Reservoir(s) :*** Which call for fixing their location and capacities (min, max) to be designed based on the supplies to be made to various users, in an optimal/acceptable manner.

(ii) ***Canals/Pipelines :*** Which need to be located in a manner to provide access to users with adequate design capacities to ensure supply of water over space/time.

(iii) ***Pumping Stations :*** To pump water from the aquifers (min, max) and /or stream courses and other water bodies. The locations have to be fixed and their capacities designed.

(iv) ***Soil Water:*** Used in a manner to reduce the demand on other sources to ensure optimal use of water from all sources.

When water is drawn from various sources for different uses, serious problems arise with respect to quality requirements for various users which impose certain restrictions on the sources to be looked at for drawing water for use and supplies from certain sources only to be made to the users. Also, certain restrictions have to be imposed on the quality of return-flows to ensure pollution-free water for the down-stream users as well as aquatic and other life.

Computer Aided Decision-Support System for Water Management

General

The decision making-process in Water Management is characterized by a typical hierarchical structure with the attendant high degree of complexity. However, for the water resources to play their due role in the socio-economic well-being, we need :

- Application of efficient methods/models for making a thorough analysis of the role and the socio-economic impacts on the development processes contemplated
- Appropriate decision-support systems for a rational and effective decision-making.

The development of suitable computer based management information systems, tools of systems analysis backed by computer networks could make it possible to provide the necessary offline/on-line decision-support system to ensure the national development process to proceed in an optimal manner.

Structure of Decision Support System

The computer-based decision-support system is diagrammatically shown in Fig. 7.3. This can be conceived to comprise the following:

- Measuring and Information System

  This includes :

  Measuring/Transmission/Storage equipment [hardware manually operated or automated]

  Data acquisition-transmission-storage process

- Analysis/Evaluation system

  This includes:

  Tools of systems analysis & user software

  Processing which yields results, report solutions

- Organization Structure
- Operation/Regulation manuals

Task Performance

In developing and implementing the above, typically the scopes and decision levels have to be conceived to have different time horizons, as for example:

1. Planning, constructing and dimensioning of water resources systems	Longer time horizon of 20 years or more
2. Management tasks in a given water resources system	Long term and medium term management
3. Operational/Forecasting/ Control tasks for real-time operation	Short term such as days/ weeks/months/ seasons, etc.

Each of the above tasks are taken into consideration in the over-all system shown in Fig. 7.3. It is obvious that the significance of the various components changes with the tasks on hand. For operational control, the problems of data acquisition/ transmission are in the foreground. The focus shifts, with the growing time horizons needed for planning/ management systems, to the tools/ software / economic criteria, etc. In any case, the uncertainty of the information to be processed increases the subjectivity in decision-making process, which cannot be completely eliminated by models.

Caution

In this context, one has to understand that the software systems that are employed are not by themselves the decision-making systems but only the decision-supporting systems. Therefore, their job cannot always be conceived to supply the decision-maker with the best solution, but to provide a basis, which ensures rational decision-making in keeping with the demands of the situation.

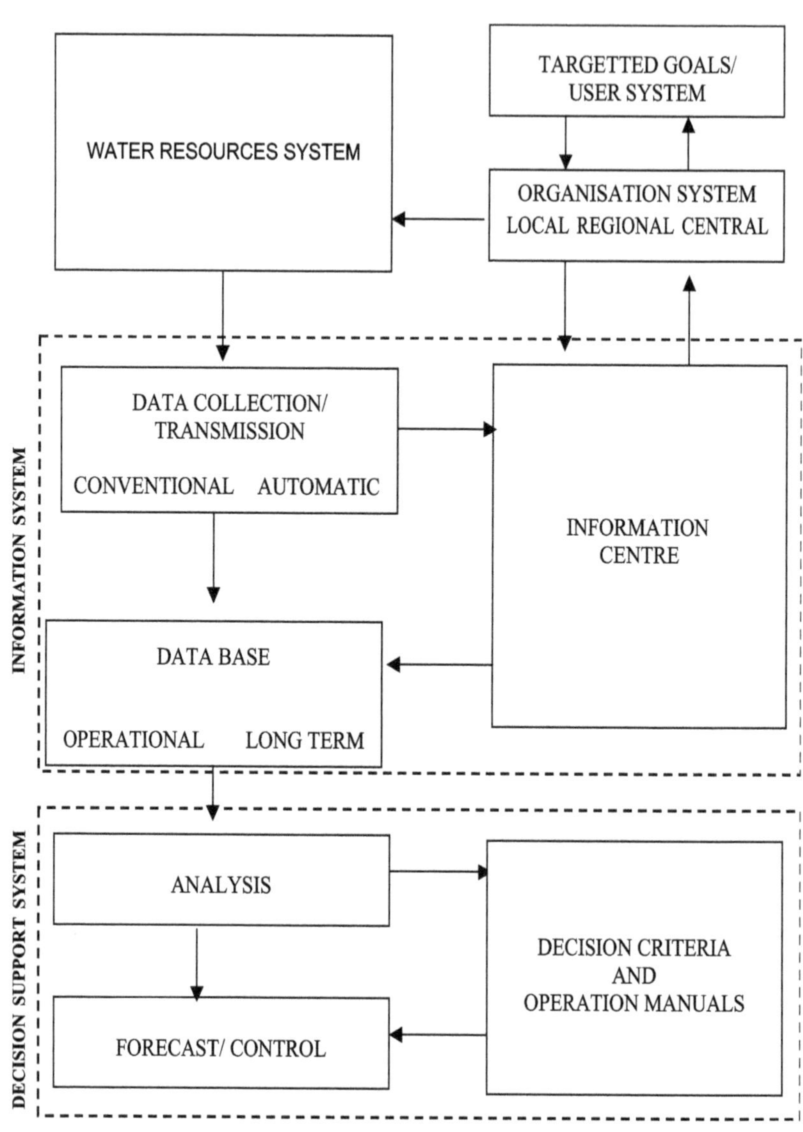

Fig. 7.3 Computer-based decision-support system.

References

Aswathanarayana, U, 2001. *Water Resources Management and the Environment*, A.A. Balkema, Lisse, The Netherlands.

Bellman, R.; 1957: *Dynamic Programming'*, Princeton University Press, Princeton, N.J.

Dantzig, G.B., 1963: *Linear Programming and Extensions*, Princeton University Press, Princeton, N.J.

Haimes, Y.Y., 1977: *Hierarchical Analysis of Water Resources Systems*, Mc Graw-Hill Book Company, New York.

Haimes, Y.Y., W.A.Hall and H.T. Freedman, 1975: *Multi – Objective Optimisation in Water Resources Systems: The Surrogate Worth Trade-off Method*, Elsevier, Amsterdam.

Hall, W.A., and Dracup, J.A., 1970: *Water Resources Systems Engineering,* Mc Graw-Hill Book Company, N.Y.

Hufschmidt, M.M. and M.B. Fiering, 1966: *Simulation Techniques for Design of Water Resources Systems*, Harvard University Press, Cambridge, Mass.

Kaden, S, Becker, A. and Gnauck, A, 1989: Decision-Support Systems for Water Management, *Proceedings of the Baltimore Symposium*, May 1989, IAHS Pub No. 180,1989.

Loucks, D.P., Stedinger, J.R., and Haith, D.A., 1981: *Water Resources Systems Planning and Analysis*, Practice-Hall Inc., Englewood Cliffs, N.J.

Ramulu, M. and Yaqoob, M., 1987: Optimal Cropping Pattern – A Case Study of Sri Ram Sagar Project, *Proc. of the Seminar on Cropping Patterns in Command Areas,* Sept 14, 1987, Water and Land Management Training and Research Institute, Hyderabad, India.0

8

Watershed Management for Drought Mitigation

[1]S.K. Jain and [2]V.P. Singh

[1]*Department of Civil and Environmental Engineering, Louisiana State University, Baton Rouge, LA 70803-6405, U. S. A.*
(e-mail : cessing@lsu.edu)

[2]*(on leave) Scientist F, National Institute of Hydrology, Roorkee, India.*

Key words : Drought alleviation, Integrated watershed management, Groundwater management, Hydrological considerations, Social and cultural considerations.

Introduction

Due to excessive and sometimes irresponsible use during the 20th century, today freshwater sources are under unprecedented stress. Many inter-related factors have contributed to this state of affairs. While the world population increased by a factor of about three during the 20th century, water withdrawals have increased by a factor of about seven due to large increases in industrial and economic activities. Consequently, one-third of the world's population lives in countries that currently experience medium to high water stress (Technical Advisory Committee, 2000). Estimates show that by 2025, about 30% of the world's population will live in countries that are already afflicted by water scarcity. Thus, even if these countries achieve higher water use efficiency and productivity, there will not be sufficient resources to satisfy domestic, agricultural, industrial, and environmental requirements. Since the agriculture sector is the major consumer of water in India, in the event of deficiency, supplies to crops will most likely be reduced, adversely affecting crop production. Pressing water needs in many countries may force people to develop additional water resources through means that may be environmentally harmful.

Increased water use has also resulted in more pollution in water bodies. Contamination of water limits the use of water in various activities as well as renders water use expensive because costly treatments are required before use. Secondly, the biological life which is dependent on water is infected by various diseases that require expensive treatments (in case of humans) and similar problems for plants and animals, including extinction (in rare cases).

The events of the year 2002 are good indicators of the water-related problems in India. Many parts of the country faced in the year 2002 the worst drought of the past 30 years. The rainfall deficiency by the middle of August was of the order of 30% or more. Consequently, municipal water supplies were badly hit in many towns and reports of crop damage came in from across the country. Even worse, large areas in the north-east were at the same time inundated by flood waters. Certainly, this was not the first time when this skewed combination of events was witnessed. We have been observing this almost every year - the only difference is in the magnitude and areal coverage. Sadly, we still don't have a workable mechanism to solve these problems. Despite more than five decades of planned development, our dependence on monsoon is not any less today.

The deficiency of water is now leading to all kinds of problems. First and foremost, it is a hindrance to economic development, and the non-availability of water and sanitation are the root cause of many epidemics. In India, it has led to riots at some places and migration of people from drought-affected areas. Worst of all, water has been the reason for many regional disputes and bitter inter-state water conflicts; many of these conflicts are still unresolved.

In order to solve the problem of drought, a long-term watershed planning and management strategy is to be evolved and implemented. This will, inter alia, require a large pool of persons with requisite training and background in hydrology and watershed management (WM). Before discussing the WM aspects, a brief description of watersheds and their modeling is given. This will shed light on the complexity of issues as well as the needs for management for drought alleviation.

A Perspective of Watersheds

The availability of water at a given place depends on the properties of its upstream watershed. Most of the consumptive water use, such as irrigation; and non-consumptive uses, such as hydropower production, recreation, and navigation, take place in the watershed itself. A watershed also receives most of the return flow from irrigation and wastewater. Due to these reasons, watersheds are considered to be the most important management units. The ecosystem approach to WM aims to integrate social, economic and environmental interests within the wider framework of the watershed. The basic motivation behind this is that people depend on their ecosystems and therefore the capacity of the ecosystems to deliver goods and services should be maintained in the long haul.

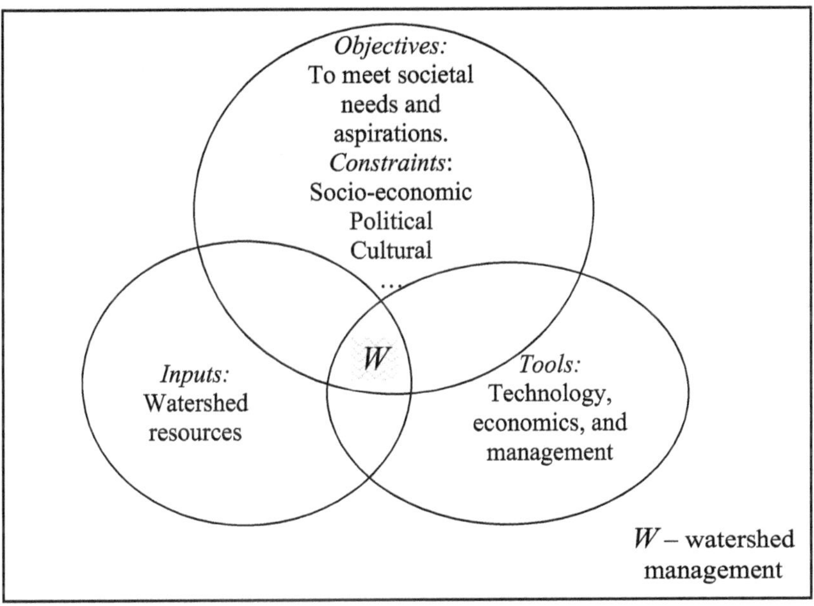

Fig. 8.1 A conceptual framework of watershed management.

The five functions of watersheds (Black, 1997) are : 1. collection of water from rainfall, snowmelt, and storage that becomes runoff, 2. storage of water, and 3. discharge of water as runoff, 4. providing diverse sites and pathways along which vital chemical reactions take place, and 5. habitat for the flora and fauna that constitute the biological elements of ecosystems. The watershed hydrologically attenuates the energy inherent in the irregular and often abrupt delivery of precipitation in a unique manner. Through outflow of water from storages, the watershed flushes a water body and this action, in turn, regulates the movement of mobilized chemicals.

A model for integrated WM helps develop implementable solutions to problems by combining all the essential component models into an umbrella scheme. The model incorporates or accumulates all of the interactive forces or influences. Hence, it aids the decision-making process and keeps the policy results within the intersection of the social goals of the management policy and the legal constraints. Such a model is shown in Fig. 8.2.

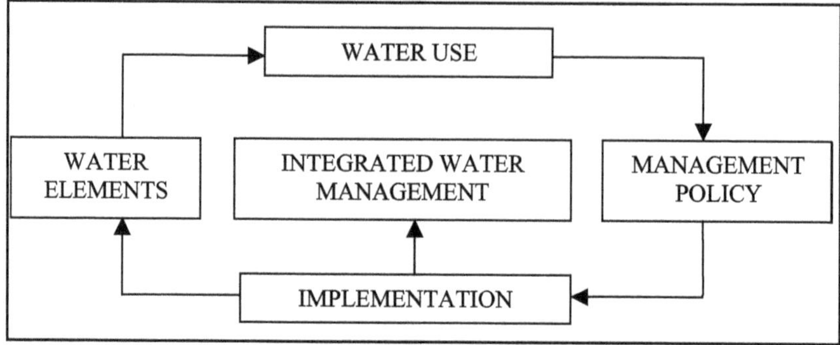

Fig. 8.2 Integrated water management.

The integrated watershed management approach is considered to have the following major objectives (Singh et al., 2002):

1. Promote sustainable economic development through
   (a) Optimum utilization of land, water, and vegetation to mitigate the adverse effects of drought.
   (b) Provide employment and local capacity building to generate income.
2. Restore ecological balance through community participation and sustainable development of natural resources, encouragement of available low cost affordable technologies for easy acceptance.
3. Improving living conditions of the poor through more equitable resources distribution and greater access to income generating activities.

Watershed modeling is an integral part of WM. Therefore, before initiating a detailed discussion of WM; a brief review of watershed modeling is in order.

Watershed Modeling

The hydrological models constitute the core in planning WM activities, in design of projects, in understanding the impact of various alternative designs, and in operation of facilities. These models also help in understanding the limits of feasible regions for various decisions and in identifying the optimal decision. Watershed hydrology deals with integration of hydrologic processes at the watershed scale to determine the watershed response. Mathematical models of watershed hydrology are employed in a wide spectrum of areas ranging from WM to engineering design (Singh, 1995). They are used in planning, design, and operation of projects to conserve water and protect its quality, planning and designing soil conservation practices, irrigation water management, wetland restoration, and water table management.

Watershed models are fundamental to WM which requires integration of watershed models with models of irrigation management, soil conservation, physical habitat, biological populations, and economic response. Estimating the value of instream water use allows recreational, ecological, and biological concerns to compete with traditional consumptive uses, i.e., agriculture, hydropower, municipality, and industry (Hickey and Diaz, 1999). Watershed models are utilized to quantify the impacts of watershed management strategies linking human activities within the watershed to water quantity and quality of the receiving stream or lake (Mankin et al., 1999; Rudra et al., 1999) for environmental and water resources protection. In summary, watershed models are an essential tool for planning, development and management of watershed resources.

A watershed hydrology model is an assemblage of mathematical descriptions of components of the hydrologic cycle. The model structure and architecture are determined by the objective for which the model is built. For example, a model for watershed planning is significantly different from the one used for water resources design or ecological management.

Likewise, a hydrologic model for flood control is quite different from the one for watershed management. Singh (1995) classified hydrologic models based on 1. process description, 2. time scale, 3. space scale, 4. techniques of solution, 5. land use, and 6. model use. An exhaustive review of watershed models and various issues related to their application has been recently provided by Singh and Woolhiser (2002). A number of models are described in Singh (1995) and Singh and Frevert (2002a, 2002b).

Integrated Watershed Models

In view of the concerns for deteriorating quality of raw water resources and impacts of climate change on water resources, the period from 1980 onwards saw attempts to link hydrologic models with those of geochemistry, environmental biology, meteorology, and climatology. This linking became possible primarily for two reasons. First, there was increased understanding of spatial variability of hydrologic processes and the role of scaling. This was essential because different processes operate at different scales and linking them to develop an integrated model is always challenging. Second, the revolution in information technology made possible the employment of GIS, remote sensing techniques and data base management systems.

Watershed-scale models that require hydrologic, crop, and economic input data and simulate the behaviour of various hydrologic, water quality, economic, or other variables under a fixed set of water allocation and infrastructure management policies are being increasingly used these days. A useful outcome of simulation of the operation of a water resources system under a range of conditions is identification of the system components that are likely to fail. The models that use detailed hydrometeorological input data can also be employed to assess the system performance under various scenarios of climate change, and changing demands, such as those due to population growth, change in command areas, cropping patterns, etc. Fig. 8.3 shows the elements of an integrated watershed model.

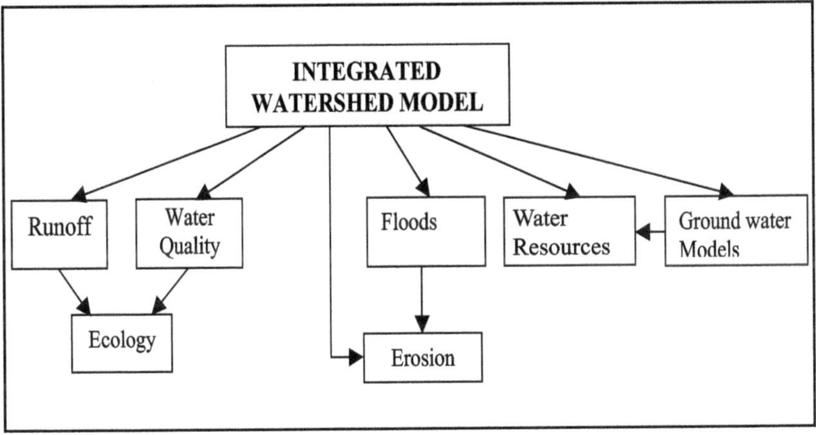

Fig. 8.3 Elements of integrated watershed model (after Harmancioglu et al., 1998).

Many of the current watershed hydrology models are comprehensive, distributed and physically based. They possess the capability to accurately simulate watershed hydrology and can be applied to address a wide range of environmental and water resources problems. 2. The scope of mathematical models is growing and the models are capable of simulating not only water quantity but also quality. 3. The technology of model calibration is much improved, although not all models have taken full advantage of it. 4. The models are becoming embedded in systems whose mission is much larger, encompassing several disciplinary areas. 5. The technology of data collection, storage, retrieval, processing and management has improved by leaps and bounds. In conjunction with literally limitless computing prowess, this technology has significantly contributed to the development of comprehensive distributed watershed models.

Watershed Planning

A sound planning is a pre-requisite for optimal use of available resources. It involves estimation of short-term and long-term needs, ways to meet these needs, and then a comparative evaluation of alternative solutions with respect to their technical, economic and social merits. Because of wide variations in distribution of resources and diversity of issues, watershed planning is always broad in scope. Such planning is needed at different levels and for different purposes. It, therefore, requires that many different uses of water, land and other resources are considered and evaluated, leading to the articulation of trade-offs among conflicting and competing objectives.

Watershed planning requires a compatible team who can draw a plan that is acceptable to the decision-maker and the public. However, future trends in demography and economy are difficult to predict and hence elements of uncertainty enter the process. The other noteworthy aspect is that many management decisions are more or less irreversible. For instance, once a dam has been built, it exists practically forever, regardless of whether there is a need for it or not. It will never be possible to restore the site to its original condition, even if a dam that is no longer needed is carefully decommissioned.

Data and Hydrological Considerations

A sound planning requires a broad database and this should include topographic maps; demographic maps; hydrogeological maps; soil and landuse maps; hydrometeorological database; water availability, demands, and water use data; and effluents discharged with and without treatment and the quality of sources of water. The agriculture-related data include cropped areas, cropping patterns, water requirements, and yields. This information would be most suitable if it is available for each planning unit, say sub-watershed. Such a database is not readily available for most watersheds in India and often considerable time is spent on creation of databases.

The use of water can be divided into two categories: consumptive use, in which water is an end to itself, and non-consumptive use, in which water is a means to an end. The consumptive use includes the use for municipal, agricultural, industrial and mining purposes. The non-consumptive uses are in-stream uses, such as hydropower, transportation, fish production, and recreation. Consumptive uses are modeled using consumptive functions and non-consumptive uses are modeled using production functions. Modern water resource projects are mostly multipurpose in nature. A drop of water may be put to a number of uses from source to final disposal as shown in Fig. 8.4.

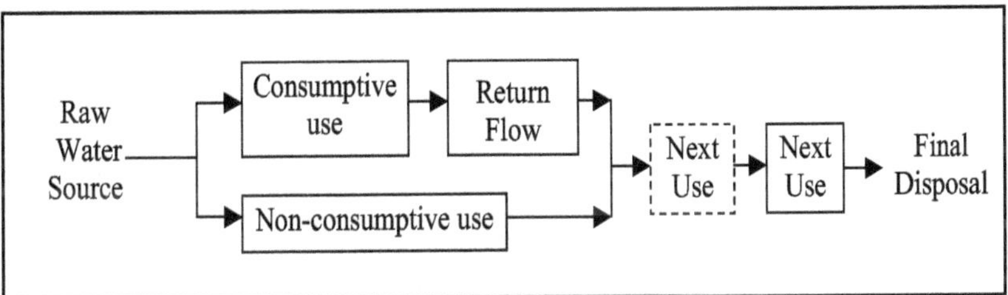

Fig. 8.4 Steps in use of water resources.

Social and Cultural Considerations

Watershed management schemes influence social, political, and cultural life of the area and it is important to consider these in the planning stage. WM activities that affect social life include building means of communication, such as roads, bridges, hospitals, parks, recreation centers, and drainage systems. As a result, trade and commerce grow, interaction with outside world grows, and there are more opportunities for social intermingling. This 'opening up' may also accelerate migration to and from the area.

The interaction with outside world also induces cultural exchange – new ideas seep in and local culture is exposed to larger audience. These may as well accelerate further WM activities as people begin to see the benefits of these actions. Sometimes, the social set-up and lifestyles change rapidly and society may undergo a complete transformation over a period of 15-20 years.

Drought Alleviation

Although India is home to about 18% of the world's human population and 15% of the livestock population, it has only 2% of the world's geographical area and 0.5% of pasturelands. The per capita availability of forests in India is about $1/10^{th}$ of the world average. In India, out of 329 million hectares of geographical area, nearly fifty percent (50%) of land is said to be either waste or degraded. Estimates show that the largest degraded agriculture land in the world is in Asia.

Clearly, the land, water, and botanical resources of India are under tremendous pressure. The carrying or sustaining capacity, in many cases, is either exceeded or is nearing the limit. This is the reason behind increasing interest in WM activities in India. The main objective of WM in such a situation is to exploit resources, such as soil, water, land, and vegetation, in a manner that these resources provide the maximum benefit to the society and at the same time are not degraded. In these settings, the key to a good WM program in India is to apply the best available technology and organize, motivate, and involve people to attain the long-term objectives of social growth and development.

Depending upon the state of development, watersheds can be classified in three groups: virgin, properly developed, and degraded. In the first case, planning will have an important role in WM while its role will be limited in the second case. In the third case, it would be desirable to place more emphasis on watershed restoration. The practices that protect, enhance, or rehabilitate the watershed are referred to as the 'best management practices' (Black, 1991).

The natural resources of watersheds have large variations and it is neither possible nor advisable to exploit all of them simultaneously; the use pattern depends upon the demand and supply as well as preferences and priorities of the society. Traditionally, water and timber are considered as the sole produce of a watershed but clearly, other products, such as forage, wild life, livestock, and recreation are also important. Many natural resources, such as water and timber, can be put to multiple uses. The multiple use management has two fundamental types: resource-oriented and area-oriented (Brooks et al., 1991). It is helpful to understand the processes and regenerative capacity of a watershed. In fact, the unbalanced exploitation and use beyond the regenerative capacity is the root cause of degradation of a large number of watersheds in India.

Current Programs

In India, watershed development or management has conventionally aimed at treating degraded lands with the help of locally available low-cost technologies (sometimes termed as appropriate technology) and through a participatory approach by close involvement of user-communities.

The broad objective of these programs was the promotion of the overall economic development and improvement of the socio-economic conditions of the poor sections of people living in problem areas. A large number of programs were formulated by the Govt. of India as well as the state governments. The Drought Prone Areas Programme (DPAP), the Desert Development Programme (DDP), and the Integrated Wasteland Development Programme (IWDP) were converted into the watershed mode in the late 1980s/1990s. Several other programs of various ministries are now being implemented with a watershed as the basic unit because the common theme has been the natural resource management for sustainable development and community empowerment.

Management of Groundwater

In India, groundwater is the main source for irrigation and drinking purposes. In early days, abstraction from the shallow aquifer was limited, mainly because demand was less and water-lifting devices were animal-powered. The abnormal growth of population and spurt in industrial activities over the last 4 decades and conversion of vast wastelands into cropping land has forced abstraction of more groundwater. The number of wells has increased and the animal-powered lifting devices have progressively replaced energized pumps.

Traditionally in India, groundwater is considered to be the personal property of the landholders. Each landholder develops his shallow or deep bore wells without any coordinated plan. The rapid development of groundwater has resulted in fast decline of water table in many parts of the country, particularly in hard rock regions, causing shallow wells to dry each year. This has resulted in deterioration of water quality and the widespread drying-up of wells following the 'failure' of the monsoon. Large resources are wasted due to drying up of wells and new wells are drilled without understanding the hydrogeological system of the basin. Many rural poor farmers are unable to deepen their wells to chase the declining water levels. This has resulted in socio-economic conflicts. In coastal areas, declining water levels are also associated with the ingress of saline water, leading to reduced crop yields, loss of drinking water supplies and ultimately loss of both fertile land and water supply wells. Though sporadic attempts have been made by to regulate the over-abstraction of groundwater, so far nothing significant could be achieved.

Given the importance of groundwater to the national economy, efficient management of groundwater resources is essential. During droughts, ground water is the only source of water at many places. Addressing the problem of over exploitation in India is a complex phenomenon. Deepening of wells does not appear to be a viable option as most wells already fully penetrate the shallow weathered aquifers. Traditional methods of groundwater management are a mix of regulation (controlling the abstraction) and pricing of the water and energy used for pumps. However, this has not worked in any part of the country.

Watershed Management for Drought Alleviation

The first step of WM is to identify the watershed and its areal extent. Selection of a watershed of an appropriate size to manage is difficult and this depends, inter alia, on the nature of the problem too. The main difficulty arises in identifying a problemshed (geographical area) that is large enough to encompass the issues but small enough to make implementation feasible. Furthermore, a watershed boundary may not always coincide with all the problemsheds. For example, problemsheds based on wildlife, forests, etc. may span over watershed boundaries. In such cases, it is advisable to establish close coordination with other organizations who have expertise in the relevant discipline. As per the Watershed Development Guidelines of the Government of India, the watershed area may be about 500 ha. This size is suitable for small-scale development and a typical NGO would be able to manage it. If the interest lies in large-scale development, one will have to adopt a much larger watershed.

Watershed and problemshed boundaries seldom coincide with political and social boundaries. Within a watershed too, many human boundaries exist, such as individual farms, villages, ethnic groups and provincial boundaries. This 'mismatch' between a watershed perspective and socio-political perspective has important implications in WM. Two options are available to identify the study area: one can choose hydrologic boundaries or political/administrative boundaries. In the first case, the organizational structure for management is based on top-level watershed, sub-watershed and so on. If the administrative boundaries coincide with the hydrologic boundaries, there is the least chance of upstream-downstream conflicts.

When following hydrologic boundaries, the entity that is responsible for overall management at the topmost level is the 'river basin authority'. In India, there are many river basin authorities, such as Bhakra Beas Management Board, Damodar Valley Corporation, Brahmaputra Board, and so on, but these cover only a small part of the country. Moreover, within the jurisdiction of their basin, these authorities control only a small domain of WM. In a large part of the country, administrative boundaries are followed for water management. As is well known, WM requires large volumes of data. Many data items that are needed for watershed management, e.g., agriculture data, are available district/tehsil-wise. If these data are to be used as in WM, these are required to be appropriately reorganized and interpolated.

The preceding discussion brings out the complexities and dimensions of WM and the interacting and involved factors. A satisfactory solution of the problem requires that one should apply the best tools that are available. Some issues, tools, and approaches that are widely practiced in the western world are being described next. These are not yet being used on a large scale in India due to various reasons, although the concepts may be well known.

Strategies for Drought Alleviation

A watershed is land area that drains water, sediments, and dissolved matter to a common outlet point. The foundation of any WM program and the activity that requires the maximum efforts is soil and water conservation. The basic idea is to arrest the rain falling from the sky at the landing place. Rather than allowing to run away and carry precious topsoil with it, water is induced to seep-in and recharge the underlying aquifer.

The minimum base flow is a commonly used statistical indicator of drought severity in a stream. The parameter that is computed for this purpose is denoted by *Q7-10* which is the lowest 7-consequtive day average flow with a probability of occurring no more than once in 10 years.

A certain percentage of this flow, e.g., 60% can be used during extreme drought situations. This parameter can also be used as an indicator of the success of WM programs.

Mapping of key watershed resources allows identification of areas that are suitable for different remedial measures. Thus, soil maps provide the basis to determine the kind of vegetation that can be planted. Overlay of thematic maps of topography, soil, and slope can help delineate areas that are most suitable for different treatments. For example, such maps are used to identify hillslopes that require treatment to reduce the velocity of run-off and check soil erosion.

Wetlands are important part of watershed management. Essentially, wetlands fall in between of dry land and water bodies and are transition between aquatic and terrestrial environments. Some familiar forms of wetlands are marshes, bogs, and swamps. In many watersheds, wetlands dry up during hot weather. For drought alleviation, wetlands that are located at suitable places in a watershed have been found to be highly effective. During storm events, wetlands slow down water movement and water is refilled in these depressions. This water gradually recharges underlying aquifer. This results in higher base flow in the stream and during drought; water can be pumped out of the aquifer by mechanical means.

Water harvesting is the only long-term viable solution to combat droughts. The most common water harvesting structure is a check dam. These are very effective in trapping rainwater and its infiltration to replenish ground water. For example, in Alwar district, Rajasthan, about 3500 johads (check dams) were built for water harvesting. The result was significant rise in water table besides making the 45 km long Arvari River perennial.

Another indigenous techniques of water harvesting in "well-in-tank" system was developed in Kutch area of Gujarat. It consists of a shallow hand-dug well, called 'virdas' at the bottom of a tank. During runoff events, the virdas are filled with water and silt. During scarcity, water of tank is used first and when the tank dries, the silt is removed from virdas and its water is used.

People's Participation

Evaluation of many past development projects has shown that poor identification of the needs of local communities and inadequate assessment of the social impacts are key reasons for project failures and nonacceptance. Local residents with deep knowledge about the area can sometimes provide better inputs than 'an outsider planner' or the officers of a governmental agency or authority may not know fully well what the public wants and what is best for them. At the same time, local communities may not be able to appreciate new ideas unless they understand the thought process which has generated these ideas. An important remedy is more rigorous pre-project analysis of social and cultural conditions, more interaction with people, consulting local communities during project design, and implementation of 'open' planning.

Public participation (PP), also termed as public or people's involvement, is the process through which the views of all interested parties (persons or groups having an interest in the project or stakeholders) are integrated into project decision-making. PP should not be limited to just ascertaining different views.

The main aim of PP is to create openness and dialogue so as to improve decision-making. In Jhabua, Madhya Pradesh, the success of WM is largely attributed to involvement and empowerment of local people. The experience showed that as government begins to sincerely work with people, particularly for management of natural resources and environment, solutions are better, have wide acceptance, and are easy to implement. When communities come together to manage these resources, the rules are strictly enforced and compliance is at a much higher degree.

Some Case Studies

A look at the current scenario of WM in India reveals that despite the odds and handicaps, many people have done praiseworthy work and have achieved success by applying the limited and conventional methods. This was possible due to their hard work, commitment, and dedication. Of course, people mostly know about only those cases that have been reported in mass media. Clearly, such cases are very few; a large number of important developments go unreported or are reported only in technical literature. It is helpful to recall a few Indian success stories.

Among the studies that have been widely reported is the pioneering work by an NGO named Tarun Bharat Sangh in Rajasthan. This NGO has taken up WM works in a number of watersheds in Alwar district and has been able to drastically improve water availability in them. They have also been involved in construction of check dams at some places with the active involvement of the local community. Another noteworthy achievement is work at Ralegam Siddhi by Sri Anna Saheb Hazare. Large scale water conservation and catchment treatments were undertaken over a large area in Maharashtra by his group and they were able to conserve water and provide irrigation to about 1200 acre (about 480 ha) land and could produce crops worth nearly Rs.5 million (eq. USD 100,000) each year. Recall that increase in agriculture output has cascading effect on many sectors of economy.

Singh et al. (2002) provide details of the noteworthy work in Jhabua district, Madhya Pradesh. The Chipko movement launched by Sri Sundar Lal Bahuguna in the hills of Uttranchal mainly to check deforestation has also generated considerable interest.

Neeru-Meeru Program of Andhra Pradesh

The total geographical area of Andhra Pradesh is 27.5 million hectares out of which waste and degraded lands constituted 11.5 million hectares (42%). The state receives an average annual rainfall of 896 mm. Out of the total volume of 244.4 x 10^9 cubic metres rainfall received, 41% is lost as evaporation and evapotranspiration, another 40% is lost as surface run-off, about 10% is retained as soil moisture, and 9% is recharged as ground water. In this scenario, the objective of the water conservation would be to develop strategies for water conservation and its sustainable utilization.

A ten-year watershed programme for development of all degraded and waste lands was launched during 1997 to treat 10 million hectares. Some time ago, all water conservation activities were brought into campaign mode to ensure convergence of the efforts of all water concerned departments in the name of *Neeru-Meeru* (Water and You) to promote water conservation on mission basis. The Neeru-Meeru concept envisages creation of awareness amongst people for ensuring their participation in water resource management.

While prioritizing areas, 494 mandals (country) were identified. Besides, villages/habitations which experience drinking water scarcity and areas with groundwater within 10 meters from surface have been included. The departments involved in executing Neeru-Meeru works include Rural Development, Forest, Minor Irrigation, Rural Water Supply, Municipal Administration and Urban Development, and Endowments. Neeru-Meeru envisages to co-ordinate and guide the water conservation efforts initiated by various departments. The activities taken up by different departments are aimed at creating more filling space for harvesting rain water which contributes to additional ground water recharge. The works involved in implementation of Neeru-Meeru programme include constructing check dams, percolation tanks, ponds, channels, and restoration of these for higher recharge of ground water.

Impact analysis studies of works taken up under Neeru-Meeru on ground water levels showed that due to deficit rainfall of 27%, the depth to water level at the end of August 2002 was 10.62 m as against 10.40 m at the end of August 2001. A net fall of 0.22 m was recorded in the state between August 2001 and August 2002, due to long dry spell and over exploitation of groundwater. Seasonal bore wells (dried up bore wells, hand pumps) reduced from 9594 to 4079 between September, 2001 and September, 2002. Drinking water transportation habitations reduced from 320 to 314 between September, 2001 and September, 2002. Expected additional recharge due to Neeru-Meeru works is 6.1 x 10^9 cubic metres.

Despite many success stories, a perceptible change in the condition of watersheds is not visible and much needs to be attained in the field of WM. Unfortunately, the country is still as vulnerable to natural water-related disasters as it was many decades ago and many WM initiatives have failed to make any significant dent in combating against droughts.

Conclusions

In view of the severity of the problems and to effectively deal with the problems, a nationwide integrated WM program must be launched. The recently launched Haryali Yojana under which all watershed management programs have been integrated is a step in the right direction. Large scale use of the emerging technology, such as remote sensing, GIS, and improved models, is necessary.

Creation of hydrologic, meteorologic, soil, and agriculture database is equally important. A database of digitized watershed maps and other hydro-met data will provide useful inputs in WM at various stages. In India, rainwater harvesting has received considerable attention during the last few years. This should lead to considerable benefits if rainwater harvesting is practiced on a large-scale. But rainwater harvesting can only partially help in extreme situations. The experience of drought of 2002 shows that so far we do not have a workable strategy and means to fight severe droughts in a meaningful way. Since preparedness is necessary to deal with these creeping disasters, WM needs to be backed up by a long-term weather forecasting set-up.

References

Black, P.E. (1991). *Watershed Hydrology*. Prentice Hall, Englewood Cliffs, New Jersey.

Black, P.E. (1997). Watershed functions. *Journal of the American Water Works Association*, 33(1), 1-11.

Brooks, K.N., P.F. Ffolliott, H.M. Gregersen, J.L. Thames (1991). *Hydrology and the management of watersheds*. Iowa State University Press, Ames.

Hickey, J. T. and Diaz, G. E. 1999. From flow to fish to dollars: An integrated approach to water allocation. *J. Am. Water Resource. Assoc.*, 35(5), 1053-1067.

Mankin, K. R., Koelliker, J. K. and Kalita, P. K. (1999). Watershed and lake water quality assessment: An integrated modeling approach. *J. Am. Water Resourc. Assoc.* 35(5), 1069-1088.

Rudra, R. P., Dickinson, W. T., Abedini, M. J. and Wall, G. J. 1999. A multi-tier approach for agricultural watershed management. *J. Am. Water Resourc. Assoc.*, 35(5), 1059-1070.

Singh, A.K., Eldho, T.I., and Prinz, D. (2002). Integrated watershed approach for combating drought in a semi-arid region of India: the case of Jhabua watershed. *Water science and Technology*, Vol. 46(6-7), 85-92.

Singh, V.P. (Ed.) (1995). *Computer Models of Watershed Hydrology*, Water Resources Publications, USA.

Singh, V.P., and Frevert, D.K. (2002a). *Mathematical models of small watershed hydrology and applications*. Water Resources Publications, USA.

Singh, V.P., and Frevert, D.K. (2002b). *Mathematical models of large watershed hydrology*. Water Resources Publications, USA.

Singh, V.P., and Woolhiser, D. A.(2002). Mathematical modeling of watershed hydrology, *Journal of Hydrologic engineering*, ASCE, 7(4), 270-292.

Technical Advisory Committee (TAC) (2000). *Integrated Water Resources Management.* Background Paper No. 4. Technical Advisory Committee (TAC) of Global Water partnership. Stockholm, Sweden.

9

Socio-Economic Impact of Drought

Purendra Prasad

Department of Socialogy, University of Hyderabad, Hyderabad - 500 045, India (nppss@uohyd.ernet.in)

Key words : Peoples' perception of drought, Biophysical dimensions, Socioeconomic dimensions, Coping with drought, Current policy measures, Mitigation paradigm.

Introduction

On the basis of Indian case histories, the chapter seeks to elucidate how through an understanding of the biophysical and socioeconomic dimensions of drought, it is possible to develop integrated techno-socio-economic measures for the mitigation of drought.

Drought in arid and semi-arid regions had been a regular climatic feature as well as a social phenomenon for centuries in India (Table 9.1).

Table 9.1 Reported drought events in India over the past 200 years

Period	Drought Years	Period	Drought Years
1801-1825	1801,4,6,12,19,25	1901-25	1901,4,5,7,11,18,20
1826-1850	1832,33,37	1926-1950	1939,41
1851-1875	1853,60,62,66,68,73	1951-1975	1951,65,66,71,72,74
1876-1900	1877,83,91,97,99	1976-2002	1977,78,79,82,83, 85,87,88,92,02.

(Source: Land Information for Drought, 2002)

People living in these regions perceive droughts as normal and expected, like death and taxes. It is a different issue altogether that what is considered normal is actually unbearable. However, considering drought as normal has tended to routinize the management of droughts. Yet the notions of crisis associated with droughts are reasserted with each incidence of drought, since factors such as water scarcity, food scarcity, etc., affect majority of the poor among the rural communities. The social and economic impacts of droughts may continue over many years, even after physical conditions returned to normalcy. Thus, people learned to cope with certainty (droughts) and uncertainty (when rain comes). The resulting tensions and anxieties associated with scarcities are not tangible. Social scientists try to figure out how people survive and organize their economic activities and social lives in circumstances of great uncertainty created by droughts.

Drought is a natural hazard. While the drought itself cannot be prevented, it is possible to minimize the damage arising from the drought through being prepared for it. The total risk (R) due to drought (in terms of number of people who are distressed due to lack of food and water, number of persons migrating, loss of income, decline in economic activity, etc.) is a function of the hazard (H), which is the probability of occurrence of drought within a specified period of time and in a given area, and Vulnerability (V) which is the degree of loss (zero – no damage, one- total loss) caused by drought. Thus, R = H x V. Thus, by minimizing the vulnerability (V), we can minimize the risk (R). Suppose, an area is hit by cholera. Every body in that area is subject to that hazard. Well-to-do persons who can afford to boil their water, and eat warm food would have reduced their vulnerability to virtually zero – in other words, they are not at risk for cholera. It is the poor who cannot afford to boil their drinking water, that are vulnerable, and therefore at risk of being infected with cholera.

The more fragile the socio-economic fabric is, the more vulnerable that society is to the adverse consequences of drought. This is starkly evident in the case of Africa. According to the estimates of the Organisation of African Unity (OAU, Addis Ababa, Ethiopia) for the period, 1980-89, the number of people who died in the sub-Saharan Africa due to drought (450,000) was 450 times more than the number of people that died due to floods (1000). This clearly shows that drought is a great killer.

What is Drought ?

Since drought has both biophysical and socioeconomic dimensions, it is defined in different ways by different groups. Droughts are typically and most commonly referred to as being the result of a significant deficiency of rainfall. According to the Indian Meteorological Department (IMD), the country is said to experience a drought year when there is overall rainfall deficiency relative to the historical long period average (LPA) and when more than 20 percent of its area affected by drought conditions. By this definition, "the year 2002 becomes the first ever all-India drought year (since 1987) with rainfall deficiency for the country as a whole accounting to 19 percent of its geographical area". The Technical Committee on Drought-Prone Area Programme (DPAP) and Desert Development Programme (DDP) of the Govt. of India, recommended the inclusion of Moisture Index as one of the indicators of the onset of drought (1994).

According to Falkenmark et al (1990), drought is said to occur when the available moisture in the root zone falls below 30 % of the water holding capacity of the soil. In hydrological terms, drought manifests itself in terms of depletion of groundwater and lowering of the water table and level of water in the reservoirs. Economists are concerned with how the shortage of water leads to loss of jobs, and economic hardship, while sociologists and anthropologists are more interested in understanding the impact of rainfall deficiency on people and their institutions.

Thus, drought means different things to different people, depending on their specific interest in, or need for, rainfall. Each of these definitions has its own dimension of understanding drought. It is important to delineate specific approach or approaches to the understanding and mitigation of drought for each region and its people within a particular country rather than aim at a macro-level generalization.

Rainfall in India is highly variable and its periodicity and distribution in space and time are irregular, uneven and unreliable. Sometimes, erratic nature of rainfall makes high percentage annually but their nature is of little use to agriculture and mankind. In spite of high rainfall in certain areas, they too suffer from drought quite often.

Drought is a creeping phenomenon - its on-set as well as its end is often difficult to identify, because they lack sharp distinction from non-drought dry spells. As Tannehil suggested in his classic study way back in 1947, "The first rainless day in a spell of fine weather contributes as much to the drought as the last day, but no one knows precisely how serious it will be, until the last dry spell has gone and the rains have come again".

More than the total amount of rainfall, it is how that rainfall is *distributed* that is critically important. For instance, Crole writing about Kurnool district in Rayalaseema region of Andhra Pradesh, pointed out that " The amount of rainfall required for the whole year for a good crop in this district seems to be about 25" (635 mm). If, however the rains were regularly distributed about 15" (380 mm) would be sufficient. The popular saying is and there can be no doubt of its correctness - that one good fall of rain about an inch and a half (~40 mm) in each fortnight after seed is sown, is sufficient for agricultural necessities. But the success of crop does not seem to depend so much on the quantity of rain which falls, as on its proper distribution over a different periods of agricultural operations and at different stages of the crop growth" (cit. in Singh Tapeshwar, 1978, p.59-60).

Another difficulty is that drought is generally viewed as a transient phenomenon. As a result, it is usually not taken seriously once the rains have returned

Any approach to the understanding of droughts cannot afford to ignore the socio-economic dimensions of drought. According to Glantz (1987,p. 44), the impacts of droughts are pervasive while there are some obvious direct effects (e.g. crop loss, drying up of bore-wells, shortage of fodder for livestock), second and third order effects (e.g. price increases, increased food imports, surges in rural-urban migration rates) are little recognized. Therefore, many of the impacts that might be uniquely attributable to drought are difficult to identify.

Any understanding of drought must take into account the complexity of socio-cultural, economic and political organizations of societies and the consequent variations in the effects of lack of adequate rainfall in the local setting. Also, a drought is necessarily to be perceived in terms of the needs of a given community a technocratic approach alone is inadequate.

Droughts have a profound and increasing impact on third world communities. Suffering and loss are not a matter of physical damage alone, they have roots in social and cultural factors. The utmost priority is to study droughts in their social and historical context since this is the dimension that is most overlooked. Sociologists study the impact of drought vis-à-vis social relationships.

Socioeconomic Dimensions of Drought in India – A Historical Review

An important insight that emerges from the classical studies of Sorokin in 1942 is that the effects of famine are highly variable and may some times be opposite for different individuals and groups. He postulated the "Law of Diversification and Polarization", which, to put it simply, states that: disaster brings out the very best and the very worst in people - it exaggerates what is already there. This conclusion is borne out by subsequent studies.

For instance, the membership of particular castes (upper castes) may provide economic and social opportunities distinct from those associated with land ownership. Secondly, coping strategies can be effective in minimizing the short-term effects of drought. However, reliance on the internal and external resource base for mitigating short-term suffering may have negative consequences for the long-term. Coping strategies, such as selling livestock at low prices, losing adapted seed stock, making charcoal and firewood by cutting down trees, and clearing cultivated land of crop residues hinder the long-term ability of households to recover from destitution. Many of the families were losing one of their most important productive assets, namely, their adapted seed stock. Various studies (Bokil, 2000; Prasad, 1998) pointed out the probable deterioration in the quality of seed stock, which has received little attention in the literature on droughts.

Sen (1981, p. 45-6), in his entitlement approach, highlights that a person's ability to command food depends in particular on his/her ownership endowments (of land, labour, etc.) and exchange entitlement mapping (although he notes that entitlements may also operate through various forms of non market processes). Distress arises among the vast majority of the poor, not so much from an actual want of food, as from a loss of wages in other words, money to buy food as a general rule is a critical issue.

Food is always available in the market, even in the worst affected districts and at the worst times. But in the case of persons who at best merely live from hand to mouth, when deprived of their means of earning wages, they starve, not from the impossibility of getting food, but for want of the necessary money to buy it.

Gray and Kevane (1993, p.171) from their study 'Sheikan', in Sudan, raise two important themes: first of all, the entitlement theory provides an appropriate framework for understanding how the drought affected the lives of rural people. Rising food prices and crop failures were the two basic elements of a classic crises of entitlements. Limited local employment opportunities necessitated the selling of assets, borrowing and migration for employment. The position of labourers was extremely tenuous. Their livestock holdings were very low, and were rendered virtually useless because of the changes in terms of trade.

Bourdieu's concept of 'field' is useful in understanding what it means to access the competing resources in scarcity regions. He defines it as 'social arena' within which struggles or maneuvers take place over specific resources or stakes and access to them. Fields are defined by the parameters which are at stake – cultural goods (life-style), housing, intellectual distinction (education), employment, land, power (politics), social class, prestige or what ever – and may be of differing degrees of specificity and concreteness. Positions stand in relationships of domination, subordination or equivalence to each other by virtue of the access they afford to the goods or resources (capital) which are at stake in the field. These goods can be principally differentiated into four categories: economic capital, social capital (various kinds of valued relations with significant others), cultural capital (primarily legitimate knowledge of one kind or another) and symbolic capital (prestige and social honour) (cit. in Jenkins, 1998).

Coping With Drought - Some Indian Cases

The worst hit regions that have been suffering from drought for past several years in India are the tribal areas of Rajasthan, west Madhya Pradesh, north Gujarat and west Orissa and the poverty-stricken parts of north Karnataka, south Tamilnadu, west Andhra Pradesh and west Maharashtra. The worst affected people are the tribals and *Dalits* (former Untouchables), who constitute majority among the landless and agricultural labourers. Among them, the women are the worst hit since they are the ones who are required to ensure regular supply of water, fodder and fuel. Since it is these items that disappear the fastest in a drought year, the whole burden gets shifted on to them. A few studies focused on social networks such as caste, kin relations that became a reliable source of help during crisis periods. Traditional support systems existed along caste lines, and often it was easier for the poor to draw upon help from the richer peasants of their own caste than from outside (Caldwell et.al, 1986, p.692). Social relationships with kin and with villagers outside the kin network served as a means of support in various ways. For instance, women borrowed small amounts of food items, fuel, fodder etc., from other women. Similarly, 'helper' relationships among men, with friends, neighbours and local kin provided reciprocal labour, a sharing of irrigation water, and the loaning and renting of agricultural implements, draft animals and machinery.

The contribution of women (agricultural labour) to household earnings was substantial as evidenced by some studies. For instance, Mencher's (1987) detailed quantitative evidence for landless and near landless agricultural labour households in 20 sample villages in Tamilnadu and Kerala brings out several important features. She found that, although the wife's earnings from agricultural work were typically half to one-third of the husband's, in absolute terms her contribution from her earnings towards household maintenance was greater than her husband's earnings. The wives typically contributed 90 to 100 percent of their earnings to the family kit, while men rarely give over 60 to 70 percent of their earnings, keeping the rest for their personal use.

A few studies that focused on social networks did not get adequate attention regarding the changes that gradually took place in marriage regulations, rituals and practices connected with it, and also how several social obligations were overlooked in crisis situations. The most common response in the severely affected drought-prone areas was seasonal migration. Region - specific studies revealed a variety of migration patterns varying by distances covered, such as circulatory migration with in a limited region, long distance movements as from Andhra Pradesh to Maharashtra; sometimes a few households went for a single agricultural operation, others for an entire season.

In drought-prone areas, most of the poor migrated in search of work. Those migrating with capital assets, such as bullocks or camel carts that were used to transport produce, were able to command higher earnings. It is noteworthy that though seasonal migration clearly helped in mitigating region-specific stress, it also depressed wages in the areas to which the migrants went and generated considerable hostility from local workers (Breman, 1985, p.335-8). Whole families were more likely to migrate among the landless than among the cultivators who tend to retain some members to take care of the farm (Agarwal, 1990, p.350). Bardhan's (1977) review indicated that migration over long distance i.e. from villages to towns/cities, features more of individual men than women and less of entire families.

Liquidation of assets, reducing the consumption patterns were few other responses that have been noticed. Considerable reduction in the possession of livestock due to distress The sale of assets during seasonal troughs, which Chambers (1981) terms 'poverty ratchets', may leave poor households worse off each year in relation to previous year to the extent that they are unable to replenish the assets or repay the loans incurred by the year end.

Variation in withstanding drought reaches down to the household level. Here the very young and the old are in greatest jeopardy. Montgomery's (1977) assessment of nutritional status among householders in South India showed that the aged are consistently ill nourished. Similarly, adequate nutrition for the young sometimes receives secondary priority among the impoverished. In drought-prone regions, the health of children deteriorated more rapidly than that of others. Mayer (1975) states that old age and its liabilities begin at about 45 years in a drought environment. Among the rural groups, the landless labourers, and marginal farmers tend to consume nutritionally inferior diet (though a shade better than urban slum dwellers). Even the level of energy - the basic nutrient, in their diets fell short of recommended dietary allowances (2400 Kcal per capita, per day), leave alone vitamins and minerals (Prahlada Rao et. al, 1987, p.1).

According to Bina Agarwal (1990, p.345), the mechanisms adopted by households for coping with seasonal shortages, fall broadly into five categories: (i) diversifying sources of income, including seasonal migration; (ii) drawing upon communal resources village common lands and forests; (iii) drawing upon social relationships patronage, kinship, friendship and informal credit networks; (iv) drawing upon household stores (of food, fuel, etc.) and (v) adjusting current consumption patterns; and drawing upon assets. These are not mutually exclusive but are typically adopted in combination.

Chen (1991, p.228-30), from her study in Maatisar village of Gujarat, finds that several coping strategies proved less viable under drought conditions than under normal seasonal conditions, notably: entering the local labour market (as the demand drops sharply under most drought conditions); drawing upon common property resources (as the competition over scarce resources from common grazing lands or private fallows increases); and drawing upon social relationships (as patronage, caste and even kin-based relationships become less reliable). Further, many households reduced consumption and took smaller but more frequent loans during the drought than they did during normal slack seasons.

Kameswara Rao (1974) through his study of 'Avancha', a Telangana village brought out the changes that occurred in the social system during the period of drought which lasted from 1971 to 1973. He finds that while the drought situation was disadvantageous to the lower sections of the society who mainly depend on manual labour, the well-to-do sections among the Kapu, Muslims, Golla and Kurava castes found it to be advantageous to them. As they were getting labour at cheaper rate, they took the opportunity to get their irrigation wells repaired and new ones dug at a much lower cost. Other findings include that of a few upper caste persons resorting to the occupations which were considered by them as demeaning; inter-marriages among different sub-castes; breaking up of joint families; reduction in the social visits; and change in the traditional service caste occupations.

Karanth's (1991, p.7-8) Chitradurga study reveals some other strategies adopted in Karnataka village of Tamlihalli: such as, growing a mix of crops and or rearing livestock; entering labour and tenancy; drawing upon the previous year's stock of grains or fixed assets, adjusting consumption, borrowing, and relying upon traditional "social security" system. He says, as may be expected, the ordering of these strategies varies among specific category of households (e.g., landless, small, marginal and big farmers).

The study by the author of the socio-economic implications of drought in his own village in Anantapur district in Andhra Pradesh, revealed the following manifestations: permanent migration by few households, increase in the area of land under cultivation, increased leasing activity among seasonal migrants, significant change in land transactions (sale and purchase), lower caste groups organizing themselves into pressure groups, which resulted in surplus land distribution, assertiveness by lower groups etc, effective utilization of common property resources (CPRs) and increase in non-farm activities. The other implications include that of children abandoning schools during migration periods which lead to large number of school drop-outs, vulnerability to ill-health and accidents at the place of migration. The pressure groups formed through migration and culture-contact curbed the power of the dominant caste, thereby modifying the dynamics of traditional relations between and among the caste groups. The study also revealed that over a period of time, people belonging to backward castes and scheduled castes improved their socio-economic condition in the long-run, while people from the upper castes became losers as 'tradition' acted as an obstacle for them. With the absence of a need-based government programme, it was found that the village poor relied on their own initiatives in adjusting to shortages (Prasad, 1998).

Sainath's study (2001) revealed that the farmers from Anantapur district took loans from the banks using the only bits of gold or jewelry they owned as collateral. Many have paid back the principal and more, but they have been crushed by the interest. Large number of people incurred debt for medical reasons too i.e. 'health loans'. Several of them were groundnut farmers owning between two and five acres. Some of the practices adopted in this district as well as in parts of Rajasthan also had its adverse impact. For instance, an unshakable faith in borewells. Often farmers with just five acres of land have invested in as many as 12 to 14 borewells. Apart from debt burden this brings, the practice has had a predictable impact on the water table. All this comes atop decades of neglect of traditional water harvesting systems, denigration of centuries old thousands of tanks etc, so the damage has been severe.

The above studies demonstrate the relation between social structure and scarcities that people encountered quite frequently in arid and semi-arid regions. Besides, malnutrition and lack of proper sanitation often lead to the spread of diseases such gastro-enteritis, malaria, tuberculosis, etc,. Thus it is evident that there is an urgent need to develop need-specific and context sensitive interventions.

Current Policy Measures and Initiatives

The drought relief measures taken up by the state include emergency distribution of water by tankers; emergency construction of wells and tube-wells; establishment of fodder supply centers; provision of rations at fixed prices; and work relief projects aimed at providing employment. Many of these projects are specifically aimed at building capital works (water storage or irrigation, for example) aimed at reducing risks in future droughts. However, serious lacunae in implementation of several of these projects and more particularly the way the ruling class and upper castes manage to garner large benefits even from the distress reduction programmes has been brought very clearly by several studies (Sainath, 1995; Mehta, 2000). For instance, while the scarcity relief manual mentions road laying as the last priority, drought relief works often concentrate on laying of roads which is more beneficial to contractors and bureaucracy than to the affected persons. Similarly, fodder distribution centres are located at a distance of 6-8 kilometers away from the village. Consequently, the small and marginal farmers as well as the landless labourers who have to survive on their daily wages, have to choose between losing their wages and losing their cattle. Often, they choose to lose the cattle.

Another important intervention by the state pertains to food-based social security which is incorporated as part of poverty reduction and drought mitigation measures. However, the available studies (Bokil, 2000) point out that the present food security system reflects two basic flaws in our policy making for agriculture and rural development. First, the problems of relatively better off sections of farmers receive far more attention in policy making than the deprivations suffered by the poor. Second, while the drought-prone areas urgently need investment and infrastructure, the emphasis remains on temporary and *ad hoc* relief measures.

Recurring droughts pose perennial problems of sustenance for the poor of any region and it is often even before the failure of crops and the establishment of relief works, that the poorest people resort to out-migration. What is most disturbing about the local level impact of drought relief is the dual implications of these programmes. First, by delaying relief these regions are maintained as pools of cheap sources of labour. A substantial proportion of migrant labourers from drought-impacted areas form the backbone of urban and public construction works. Secondly, the provisioning of relief, its declaration, its disbursement, selection of work project etc, contribute towards politicizing rural economies. Persons from upper castes who are politically dominant and economically strong garner a large proportion of the benefits. In many Indian villages, it is the rich and powerful caste groups who are assured of canal water. The poor might be considered to be users, but access to this water is linked with symbolic and material forms of capital that the poor might not possess (Bourdieu, 1977).

Suggested Paradigm for the Mitigation of Drought

Falkenmark et al. (1990) emphasize that droughts and famines should not be considered as disaster *events*, but *processes*. This statement has a profound implication – if the drought is the end result of a process, we know it is coming, and we can be prepared for it. It should indeed be possible to take advance steps to mitigate its potential adverse consequences.

Droughts should not be treated as sudden "disasters", such as a rail accident, to which the government and its agencies must react and manage at the moment. It is essential to understand the concept of drought in its totality and the significance of different approaches in dealing with drought. Sociological perspectives in conjunction with the available studies in India do provide clues in analyzing the complexity of human factors and the conditions that produce persistent suffering for the vast majority of the poor in drought-prone regions.

As is evident in India, it is not so much the physical factors (rainfall, evaporation, wind etc,.) that causes calamities and consequent suffering, as it is the policies of the state (undue deforestation, irrational water distribution mechanisms, etc) which lead to a specific use pattern of the resources that proved distressful to the large majority of the communities. Hence, it is essential to recognize the serious implications of drought on human lives and reformulate the policies for a just and humane social order. Since the scarcity of water is the crux of drought, it is by taking techno-socio-economic measures to address the core issue that the vulnerability is reduced and the risk minimized. For instance, such policies should include – retrieving local scientific practices that encourage ground water recharging (rainwater harvesting, farm ponds, check-dams, dug-out ponds, rock fill dams, contour bunding, micro-watersheds, de-silting village tanks, etc.); strict control regarding over-exploitation of surface and ground water by all the concerned (including industry); encourage appropriate technologies which are cost-effective, socially adaptable (drip irrigation, sprinklers etc); agricultural practices need to be shifted from high to low water intensive crops (but commercially viable crops need to be suggested for arid regions rather than food crops alone); equitable redistribution of available water particularly river water; ensure fair price to all agricultural products by strengthening marketing network; and adequate supply of electricity during times of drought.

A policy framework that is larger and long-term must be used to prevent the onset and perpetuation of drought and scarcity conditions. In terms of drought policies, emphasis should be on recognizing the risk to livelihoods that a majority people face. It is only by contextualising and linking the physical condition of an area with the economic and social factors that a comprehensive picture of any area can be made. Though the areas that are subject to recurring droughts may share some biophysical characteristics in common, their historical, politico-economic and social characteristics form the bedrock of the life conditions in each area. It is these factors which explain why life in the drought-prone areas of Palanpur in Bihar, Kalahandi in Orissa and Kutch in Gujarat, Anantapur and Mahaboobnagar in Andhra Pradesh are different from life in the dry belts of Maharashtra and Karnataka.

The problem faced by most of the villagers in drought-prone regions is not that they resist modernization or new urban influences; they are in fact very receptive to changing their life-styles when they believed that it would benefit them. The tragedy of their life-style was that the villagers were never able to break away from a crushing environment of *economic and caste* poverty that frustrated their every attempt to improve their socio-economic condition (Freeman, 1977). Unless a comprehensive and a well-thought out approach is adopted for drought, the victims will continue to live, not with droughts but the politics of it.

References

Agarwal, Bina (1990). Social Security and the Family : Coping with Seasonality and Calamity in Rural India. *The Journal of Peasant Studies*, 17(3): 341-412.

Bardhan, Kalpana (1977). Rural Employment Wages and Labour Markets in India – A Survey of Research III. *Economic and Political Weekly*, xii (28): 1101-1118.

Bokil, Milind (2000). Drought in Rajasthan. *Economic and Political Weekly*, 25 November.

Bourdieu, Perrie (1977). *Outline of a Theory of Practice.* Translated by Richard Nice, Cambridge: Cambridge University Press.

Breman, Jan (1985). *Of peasants, Migrants and Paupers*. Delhi: Oxford University Press

Caldwell, J.C, P.H Reddy and P Caldwell (1986). Periodic High Risk as a cause of Fertility Decline in a changing Rural Environment : Survival Strategies in the 1980-83 South Indian Drought. *Economic Development and Cultural Change*, 34: 677-701.

Chambers, Robert (1981). Introduction In R Chambers, R Longhurst, and A Pacey (eds). *Seasonal Dimensions of Rural Poverty*. London: Frances Pinter.

Falkenmark, M., Lundquist,J., and Widstrand, C. (1990). *Water Scarcity – An Ultimate Constraint in Third World Development*, Tema V, Report 14, and Sweden: University of Linkoping.

Freeman, M.J. (1977) *Scarcity and opportunity in a dry village*. London: Cummings

Glantz, Michael (1987). Drought and Economic Development in Sub-Saharan Africa. In

Glantz Michael (ed) *Drought and Hunger in Africa*. Cambridge: Cambridge University Press.

Government of India (1994). *Technical Committee on DPAP and DDP Report.*

Gray, L., and Michael, K. (1993) For whom is the rural economy resilient ? Initial effects of drought in western Sudan. *Development and Change*, v. 24, p.159 - 176

Jenkins, Richard (1998). *Key Sociologists* – Pierre Bourdieu. London: Routledge.

Karanth G.K (1991). *Farmers Survival Strategies in a Drought-Prone District – A Case Study of a Village in Chitradurga District.* Bangalore: ADRT Unit, Institute for Social and Economic Change (Mimeo).

Mehta, Lyla (2000). Drought Diagnosis – Dryland Blindness of Planners. *Economic and Political Weekly*, 1 July.

Mencher, P Joan (1987). *Agriculture and Social Structure in Tamil Nadu.* New Delhi: Allied Publishers Private Limited.

Montgomery, Edward (1977). Social Structuring of Nutrition in Southern India. In Lawrence Green (ed). *Malnutrition, Behaviour and Social Orgnisation.* New York: Academic Press.

Mayer, Jean (1975). Management of Famine Relief. *Science*, 188:571-78.

Prasad, Purendra (1998). *Famines and Droughts: Survival Strategies.* Jaipur: Rawat Publications.

Rao, Prahlada N, D Hanumantha Rao and B S Narsinga Rao (1987). Drought and *Nutrition. Nutrition News*, 8(6), Hyderabad: National Institute of Nutrition.

Chen, Martha Alter (1991). *Coping with Seasonality and Drought.* New Delhi: Sage Publications.

Rao, Kameswara N V (1974). Impact of Drought on Social System in a Telangana Village. *Eastern Anthropologist.* 27(4): 299-314.

Sainath P (1996). *Everybody Loves a Good Drought* : Stories from India's Poorest Districts. New Delhi: Penguin.

Sainath, P. (2001) Where stomachs are terminals. *The Hindu*, 29 April, 2001

Sen, A K (1981). *Poverty and Famines: An Essay on Entitlement and Deprivation.* Delhi: Oxford University Press.

Singh Tapeshwar (1978). *Drought-Prone Areas in India.* New Delhi: People's Publishing House.

10

Aberrated Climate Systems and their Impact on Water Resources

A.A.L.N. Sarma

Department of Meteorology & Oceanography, Andhra University, Visakhapatnam – 530 003.
E-mail: aalnsarma_met@rediffmail.com

Key words : Aberrated climate systems. Forcings. Climate Change projections. Impacts on Hydrological cycle. River flows. ENSO and LNSO. Challenges ahead

Introduction

Water use has been growing at more than twice the rate of population increase and a number of regions of the world are chronically short of water. The spatiotemporal distribution of available fresh water has a bearing on the state of the world climate system and is replenishing or depleting the global water system projects across the world climate spectrum. Further more the state of the world climate system has been subjected to

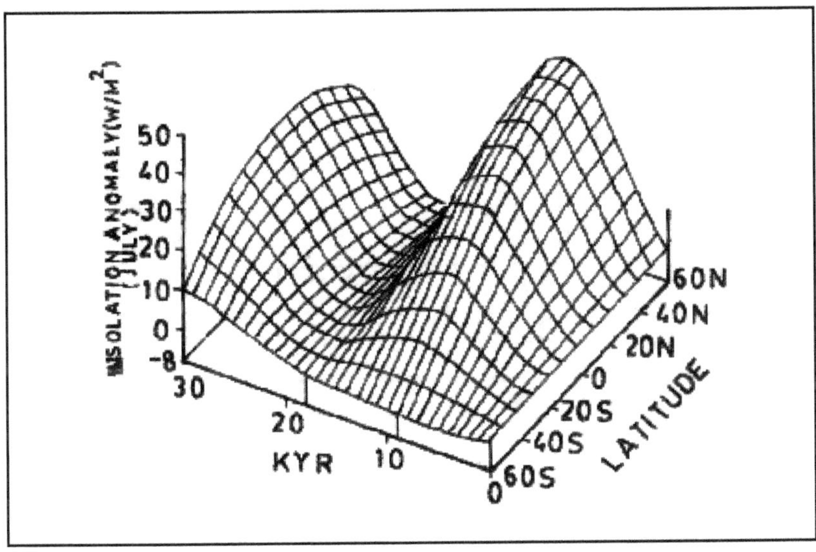

Fig. 10.4 Changes in solar insolation over time and latitude are related to variations in the orbital geometry.

Source : Nicklas and Imbrie, (1986/87) Oceanus.

The signal of orbital variations is clearly seen in the Quaternary climatic record from deep sea sediment cores. For instance, the relative proportion of heavy (^{18}O) to light (^{16}O) isotopes of oxygen recorded in the carbonate skeletons of plankton (which are preserved in deep sea sediments) varies in response to long term climate variations with heavy oxygen being relatively concentrated during the intervals when the ice sheets are larger and oceans are cooler. It is hence possible to delineate the existence of alternating cold and warm periods, with warm relatively ice free inter glacial periods, occurring every 100,000 years or so and lasting for about 10,000 years.

Glacial periods are initiated by reduced winter summer contrast (due obliquity and /or a precessional position such that perihelion coinciding with the equinox). The resulting mild winters will enhance the

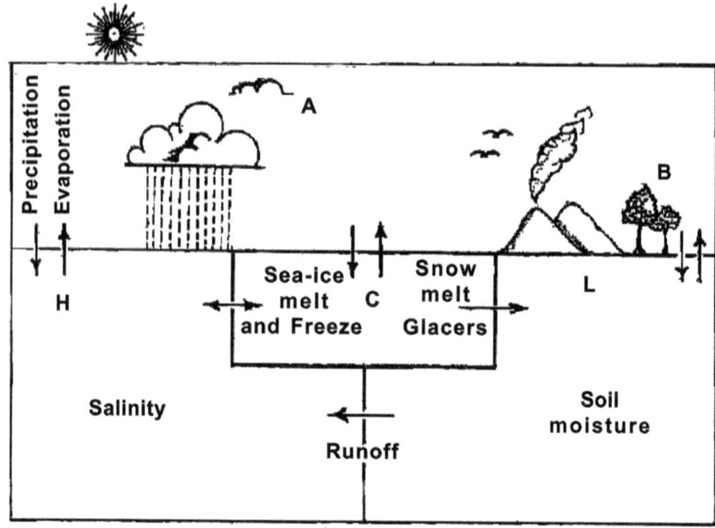

A : Atmosphere ; H : Hydrosphere (Ocean) B : Biosphere ; L : Lithosphere (Land)
C : Cryosphere (Snow & Inc)

Fig. 10.1

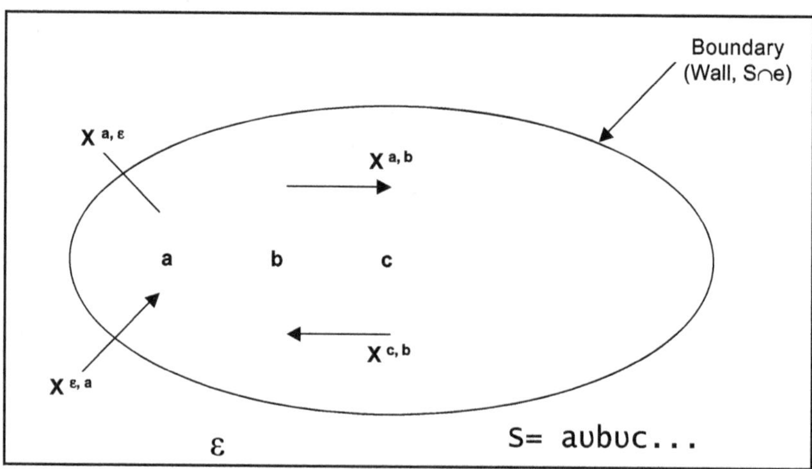

Fig. 10.2 Thermodynamic system with the subsystems and their interactions among them and environment.

Source: Peixito and Oort (1993), Physics of climate.

among the subsystems. Part of the output goes to as input, so that the net response of the system is altered; the feedback mechanism act either to enhance the final output (positive feed back) or to dampen it (negative feedback). Under equilibrium conditions the net radiation (solar minus terrestrial) at the top of the atmosphere (F_{TA}) is zero. The mean surface temperature of the earth (T_S) is regarded as one of the responses of the climate system to the imposed external forcings with all possible internal feedbacks. Any external perturbations, due to changes in solar output, atmospheric water vapour, clouds, carbon dioxide, volcanic eruptions, etc. will set an imbalance in the net radiation at the top of the atmosphere (ΔF_{TA}). If the system is damped, the variables of the climate system including T_{sfc} will change in order to adjust to new equilibrium state. So, consider ΔF_{TA} as the signal input for the climate system and ΔT_{sfc} as the output (Peixito and Oort, 1993). So that

$$\Delta T_{sfc} = G_F \ \Delta F_{TA} \qquad \text{.....(10.1)}$$

Where, G_F is a sensitivity factor that includes all feedback processes and is denoted as the gain. For the feedback mechanisms the input signal is now ΔT_{sfc} and each of the process (albedo, water vapour, lapse rate, clouds and CO_2) is characterized by a transfer function, H_i. Assuming that the feed back processes are independent we can write for the final input

$$\Delta F\hat{e}_{TA} = \Delta F_{TA} + (\Sigma H_i)\ \Delta T_{sfc} \qquad \text{.....(10.2)}$$

Added to the external forcing ΔF_{TA} is a net effect by the second item on the R.H.S of eq. (10.2) resulting from internal feedback processes, acting by themselves. Then we can write

$$\Delta T_{sfc} = (G/1\text{-}G\ \Sigma H_i)\ \Delta F_{TA} = (G/1\text{-}\Sigma f_i)\ \Delta F_{TA} \qquad \text{.....(10.3)}$$

Where G is the gain without feedback and f_i represents the feedbacks of the different processes. This type of analyses or some other variations of it are being used in assessing the behaviour of numerical climate models.

Forcings

Among the five subsystems of the world climate the subsystem of the atmosphere – the fastest responding and most variable – is most directly responsible for the present climate conditions, but atmospheric conditions are also dependent on the temperature and circulation patterns of the ocean surfaces and on the extent of major ice sheets. The slow response components of the oceans and cryosphere have a damping effect upon the fast response once. The changes in the earth's orbital forcing (Fig. 10.3, source: Nicklas & Imbrie, 1986/87)), with amplification from greenhouse gases (GHG's) (Table 10.1 : source: IPCC, 1996), are largely responsible for a variety of phenomena, such as, the long- term trend in ice sheet growth and decay, fluctuations in the strength of the tropical monsoon systems (lasting of the order of 20,000 years), glacial and inter glacial fluctuations (with a periodicity of about 100,000 years), and long term fluctuations of several million years in duration. These are manifested in the form of land -sea disposition, ice cover and resulting changes in albedo (a ratio of reflected to incident solar radiation) and sea level.

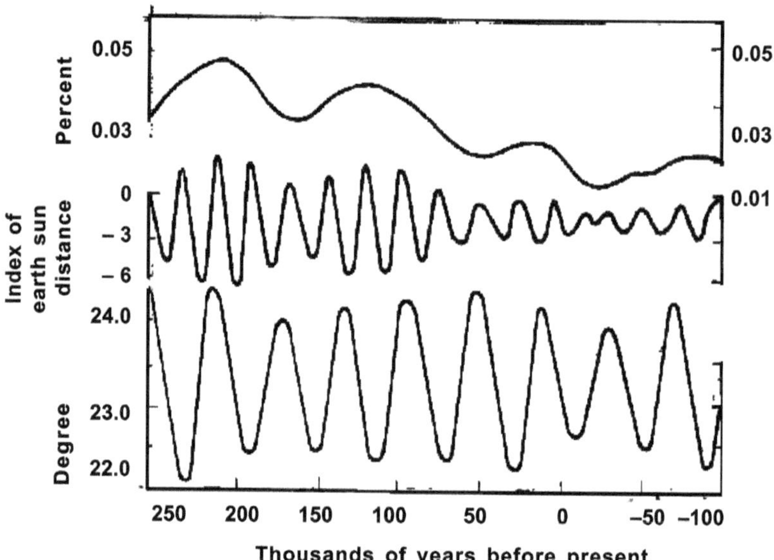

Fig. 10.3

Table 10.1 Greenhouse gases – Human activities

	CO_2	CH_4	N_2O	CFC -11	HCFC -22	CF_4
Pre-Industrial Concentration	280 ppmv	700 ppbv	275 ppbv	0	0	0
Concentration In 1994	358 ppmv	1720 ppbv	312 ppbv	268 pptv	110 pptv	72 pptv
Rate of change in concentration	1.5 ppmv/yr 0.4%/yr	10 ppbv/yr 0.6%/yr	0.8 ppbv/yr 0.25%/yr	0 pptv/yr 0%/yr	5 pptv/yr 5%/yr	1.2 pptv/yr 2%/yr
Atmospheric Life time (years)	50-200	12	120	50	12	50000

Source : IPCC, (1996a) Climate change 1995 – The science of climate change.

The orbital forcing of the earth vary systematically due to changes in the eccentricity of the earth's orbit around the sun, obliquity of the earth's axis relative to the ecliptic and the precession of the equinoxes relative to perihelion (time when the earth is close to the sun) and aphelion (when the earth is far away from the sun). These regular variations in orbital geometry control the amount and the seasonal distribution of the solar radiation intercepted at top of the earth's atmosphere. For example 10,000 years ago, the tilt of the earth's axes was about 24.2 ° (as against 23.4 ° at present), and perihelion occurred in July (as opposed to January at present). Consequently the northern latitudes received about 8% more radiation in July and about 10% less in January, than at present (Fig. 10.4, source : Nicklas & Imbrie, 1986/87). The general circulation of the atmosphere and oceans had to respond to such changes in energy input.

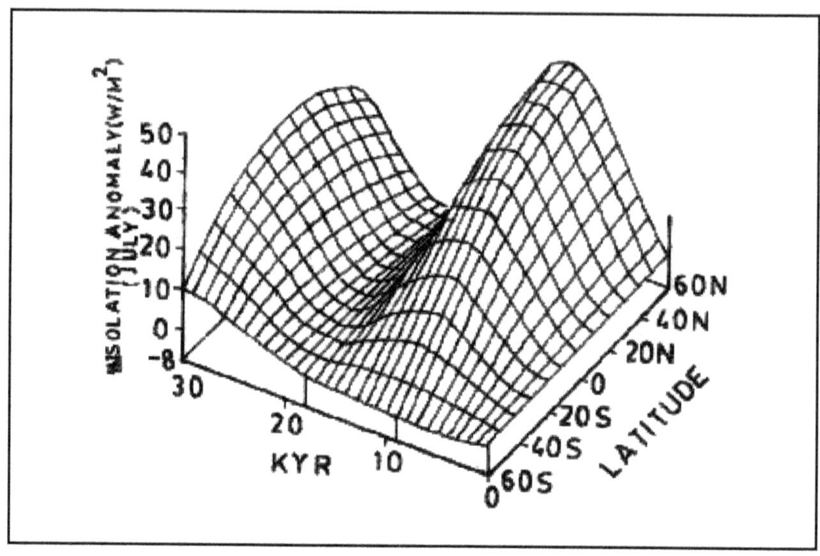

Fig. 10.4 Changes in solar insolation over time and latitude are related to variations in the orbital geometry.

Source : Nicklas and Imbrie, (1986/87) Oceanus.

The signal of orbital variations is clearly seen in the Quaternary climatic record from deep sea sediment cores. For instance, the relative proportion of heavy (^{18}O) to light (^{16}O) isotopes of oxygen recorded in the carbonate skeletons of plankton (which are preserved in deep sea sediments) varies in response to long term climate variations with heavy oxygen being relatively concentrated during the intervals when the ice sheets are larger and oceans are cooler. It is hence possible to delineate the existence of alternating cold and warm periods, with warm relatively ice free inter glacial periods, occurring every 100,000 years or so and lasting for about 10,000 years.

Glacial periods are initiated by reduced winter summer contrast (due obliquity and /or a precessional position such that perihelion coinciding with the equinox). The resulting mild winters will enhance the winter precipitations (snow) in the high latitudes of the northern hemisphere; the cool summers prevent snowmelt in the same regions. The result is glacierization. Interglacial and Interstadial periods are initiated by the opposite conditions, i.e. cold winters and warm summers. For example, the last deglaciation during 18,000 to 6,000 years, coincided with a precessional position (between 13,000 and 7,000 years ago) in which perihelion occurred during the northern hemisphere summer. Northern Hemisphere summers were therefore generally warmer than those of today.

During the past 20,000 years the climate of earth has changed from the extreme of a glacial interval to an interglacial one with insolation variations and icesheets and oceans provided a set of changing boundary conditions. Kutzbatch and Guetter (1984) reported from simulation studies that orbital changes and land surface changes in terms of the differences in input vis-à-vis the present, account for all episodes, such as 18,000 and 6,000 years ago as well as 9,000 years ago. The results suggest that the earth's climate was colder and dryer 18,000 years ago than now, with a similar seasonal contrast between summer and winter. The simulation for 6,000 years ago suggest that the climate has warmed markedly from the conditions of 18,000 years ago and the range in seasonal temperature was larger. The present Holocene interglacial changes whose length is around 10,000 years must be attributed to variations of the hydrologic cycle since the continental ice sheet has disappeared and is dwarfing those of the last couple of centuries.

The aforesaid evidences are of importance in understanding the impact of climate variations on human activities during the past 10,000 years because they contributed not only to the rise of the temperature of specific geographical realms but also on the globe as a whole (Fig. 10.5, source : IPCC, 2001).

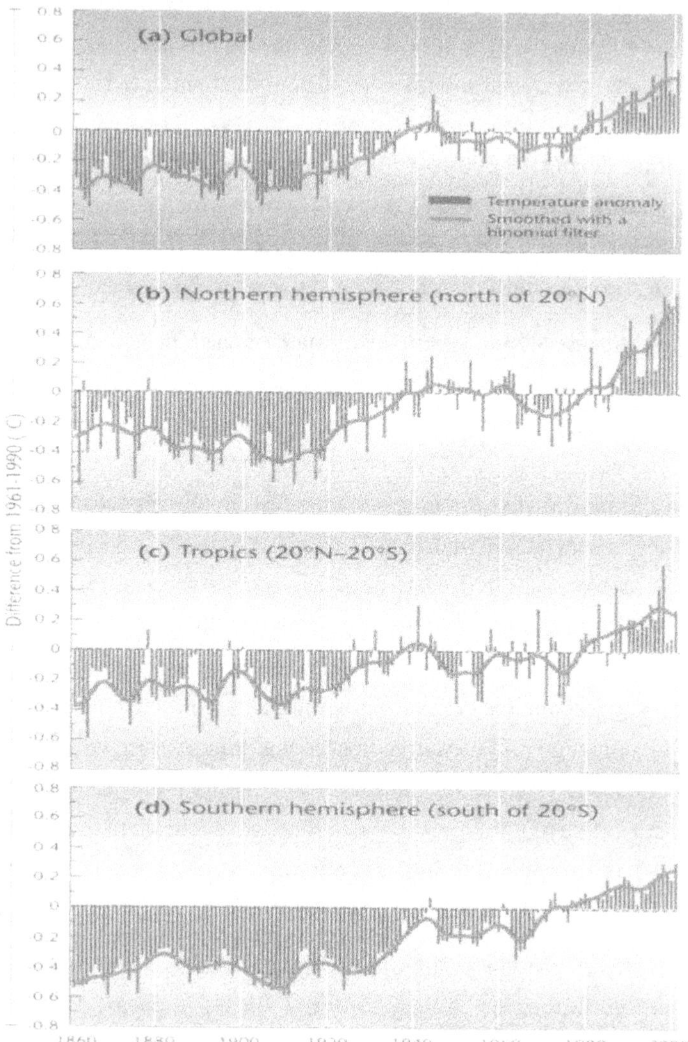

Fig. 10.5 Combines annual land-surface air and sea surface temperatures from 1861-2001 relative to 1961-1990.

Source : IPCC, 2001 and Climatic Research Unit and Hadley Centre, The Met Office.

The changes during the 20th century in weather and climate patterns at global as well as hemispheric scales, might be due to the acquired radiative forcing of the climate system. Clearly many of the changes during the 20th century climate extremities are related to increase in global and hemispheric mean temperatures accompanied by mean sea level rise and lengthened growing season. They are tied to increased concentrations of the Green House Gases (GHGs) (Table 10.2, source: IPCC, 2001). In the absence of climate policy interventions, the consequences of climate change are projected till the end of the 21st century. The consequences of climate change projections in terms of heat related deaths, floods and storms, loss of coastal wetlands, decrease in crop yields in tropical and sub tropical regions and water demands are projected to increase further. It is interesting to note that the middle and high latitudes would be benefiting from climate change till 2025 (Table 10.3, source: IPCC, 2001).

The constant circulation of water in various phases, which plays a crucial role in climate system of the earth, is powered by direct and indirect influence of solar energy. It constitutes the hydrological cycle and is active where energy and moisture are plenty, with limiting factor being either energy or moisture. The hydrologic cycle has two major branches – the terrestrial and the atmospheric. The two branches of hydrological cycle join at the interface between the atmosphere and the earth's surface. Fig. 10.6 (source: *Scientific American*, 1989) schematically depict the distribution and the flow of water in the climate system.

It is very interesting to note that the loss or "output" of water from the earth's surface through the evapotranspiration constitutes the "input" of water for the atmospheric branch, where as precipitation is the atmospheric output and constitutes the gain for the terrestrial branch of the hydrologic cycle. Globally averaged precipitation is 2.4 mm per day. The interannual variability and the trend in the global precipitation are shown diagrammatically in Fig. 10.7 (source: GEWEX project).

Table 10.2 Greenhouse gases – Human activities

Indicator	CO_2 Ppm	CH_4 ppb	N_2O Ppb	Tropo-spheric O_3	Global Ts	NH Ts	Diurnal Range	Global MSL	Growing season
Observed Change	228-368	700-1750	270-316	35%	0.6%	Increased	Decreased	Increased at an annual rate of 1.2 mm during 20th Century.	Lengthened by 1-4 days.

Table 10.3 Climate change - Absence of climate policy intervention

Year	CO_2 Ppm	Global T (0C)	Global MSL (cm)	Heat Stress	Floods and Storms	Coastal wetlands	Crop yields	Water supply	Water demand
2025	405-460	0.4-1.1	3-14	Increase heat related deaths and illness	Increases	Loss of coastal wetlands to sea level rise. Increased erosion of shore lines.	Increase in mid and high lats. Decrease in tropical and sub-tropical regions.	Peal flow shifts from spring to winter where snowfall is a source of water.	High temps will tend to increase demand.
2050	445-640	0.8-2.6	5-32	May amplify	Greater increase	More extensive	Mixed effects in mid and high lats. Pronounced decrease in tropical and sub-tropical lats.	Water supply decrease in water stressed countries.	Water demand effects amplify.
2100	540-570	1.4-5.8	9-88	May amplify	Further increase	Further loss	General reduction in mid lats.	Water supply effects may amplify.	Water demand may amplify.

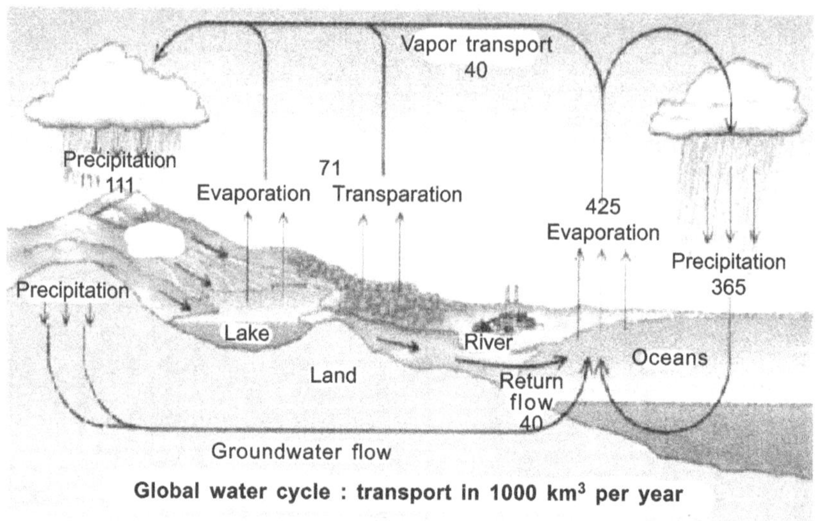

Fig. 10.6 Global hydrological cycle.

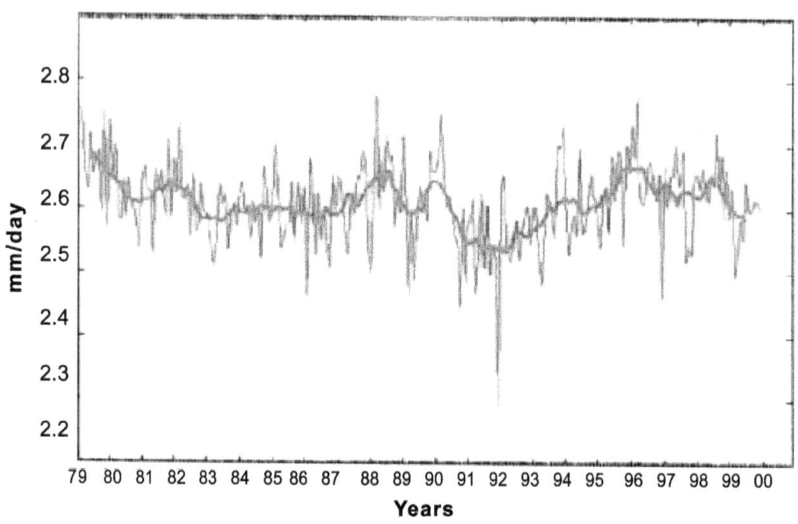

Fig. 10.7 Match of daily global precipitation.

Source : GEWEX global precipitation climatology project.
NASA/GSFC : NOAA/NESDIS : NOAA/NCEP.

It is reported since the year 1975 the short-term climate signal El Nino – Southern oscillation (ENSO) is frequently visiting and disturbing the global precipitation belts both in amount and duration, thereby triggering droughts and floods in particular locations across the world. This short-term climate signal was the outcome of Tropical Ocean Global Atmospheric Experiment (TOGA). It is related to aperiodicity from southern tropical Pacific ocean and has warm and cold phases, termed as ENSO and LaNina Southern oscillation (LNSO). It is now generally accepted that a bad monsoon occurs when southwest monsoon is in phase with ENSO, and a good monsoon is likely when it is in phase with the LNSO. The impacts of ENSO and LNSO on the daily global precipitation are clear from the Fig. 10.7 source : GEWEX Project). The years 1988 and 1992 recorded the highest (2.78mm) and the lowest (2.29mm) precipitation respectively.

Ascertaining the climate status of the land areas of the planet earth through the revised water balance concept of Thornthwaite and Mather (1955) serves as one of the techniques in understanding the hydrologic regime. The long-term balance of the water over the whole world is synonymous with the hydrologic cycle. Gone are the days when the sum total of all available water remains unchanged. This is so because the variations in anthropogenic forcing have already perturbed the energy balance of the earth–atmospheric system, thereby intensifying the hydrologic cycle. So, from the water balance concept one can account for not only all the water along with the anomalies over the globe on annual basis but also its spatiotemporal variations over the globe. The principle elements of the water budget are precipitation, potential/actual evapotranspiration (combined water loss from soil and vegetation under unlimited water supply at the root zone of the soil) and runoff. It is worth mentioning here that precipitation and evapotranspiration result from different processes and are often not in consonance with one another in terms of amount or distribution through the year at a given place. In some regions more precipitation occurs consecutively month after month, and the excess recharges the soil horizon or moves through the upper soil layers to feed streams and rivers and ultimately ends in the oceans. In other areas there is less precipitation month after month than is required by the vegetation. The soil dries out and a moisture deficit builds up and consequently there would be no run off except locally where the soil cannot absorb the rains fast enough .As a result there is perennial river flow, and drainage to the ocean is zero. In most areas of the world, however, the precipitation is deficit in one period and excessive in another so that a dry period is followed by one with a surplus and runoff. The march of precipitation through the year is never in agreement with the changing demands of the water in any part of the world. So, we need to be familiar with the occurrence and amount of such hydrologic elements (such as water surplus and deficit both locally and worldwide) to understand the role of water budget in determining the quality and quantity of stream flow, in the pattern of vegetation, and the scheduling of irrigation, etc.

We must also recognise the relation of water budget factor to other environmental factors since both willful and inadvertent changes in the physical environment have profound effect on the factors of water budget. Normal annual totals of precipitation, potential and actual evapotranspiration and surplus are diagrammed in an attempt to understand the latitudinal variation of water budget elements from $80^0 - 85^0$ N to $55^0 - 60^0$ S through the equator for land areas of the globe (Fig. 10.8, source: Legates and Mather, 1992). It is observed that there is a marked decrease in precipitation in both the hemispheres from the equatorial belt of high precipitation (10^0 S to 10^0 N) to the subtropical high-pressure belt (25^0 to 35 N and S). In both the hemispheres precipitation returned to a secondary maximum between 50^0 N – 55^0 N and 40^0 S – 45^0S with 531mm and 913 mm respectively. The latitudinal belts of 10^0 N and 20^0 N and between 15^0 S and 20^0 S recorded the maximum water excess owing to cloud cover that develops over the Inter Tropical Convergence Zone (ITCZ) between 10^0 N and S. The secondary maximum values are reflected in cases of actual evapotranspiration and water deficit also. Maximum water surplus equivalent to stream runoff occurs between 5^0 N and S concurrent with the band of maximum precipitation, with a rapid decrease poleward in both hemispheres (20^0 S and 35^0 S and 30^0 N to 35^0 N). In both the hemispheres the secondary maximum streamflow peaks occur as evapotranspiration decreases relative to precipitation. The intertropical zone is characterised by the low water deficit where as 15^0 N and 25^0 N that engulf the great deserts of the globe records the maximum.

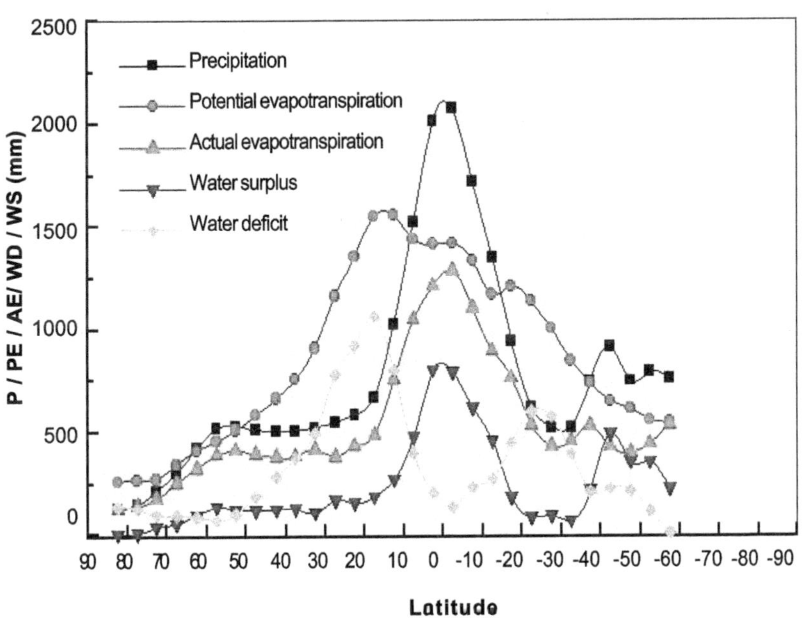

Fig. 10.8 Latitudinal annual water budget elements - Land areas.
Source : Legates and Mather (1992), Geogr. Rev.

The positive radiative force of the world climate system causes changes in the circulation pattern associated with the ITCZ and Polar front with respect to their geographical realms that give rise anomalies in terms of abnormal or subnormal hydrologic fluxes over the land areas of the globe. One of the most important attribution issues for changes in climate and weather extremes relate to the theory of radiative forcing of the climate system, which might be responsible for the intensification of the hydrologic cycle as global warming progresses. Trenberth (1996) provides a conceptual model (Fig. 10.9) for a theory to understand the following changes related to the intensification of the hydrological cycle:

- An increase in evaporation from the tropics (IPCC, 1996).
- Increase in convective clouds and related cirrus (IPCC, 1996).
- Increased continental cloud cover contributing to reduced diurnal temperature range and reduced evaporation from water surfaces over land (IPCC,1996)
- Increased precipitation in middle and high latitude land areas as contributing to enhanced evaporation and more runoff.
- Increased atmospheric water vapour over North America, China and tropical regions (IPCC,1996)
- Increased precipitation intensity in many portions of Northern hemisphere (Karl and Knight, 1997).
- An increase in the severity of tropical storms (Lambert, 1996).

Many of the above changes have been projected to occur as global temperatures increase due to increases in atmospheric GHGs, resulting in increased precipitation intensity, more precipitation in middle and high latitudes and increased atmospheric water vapour.

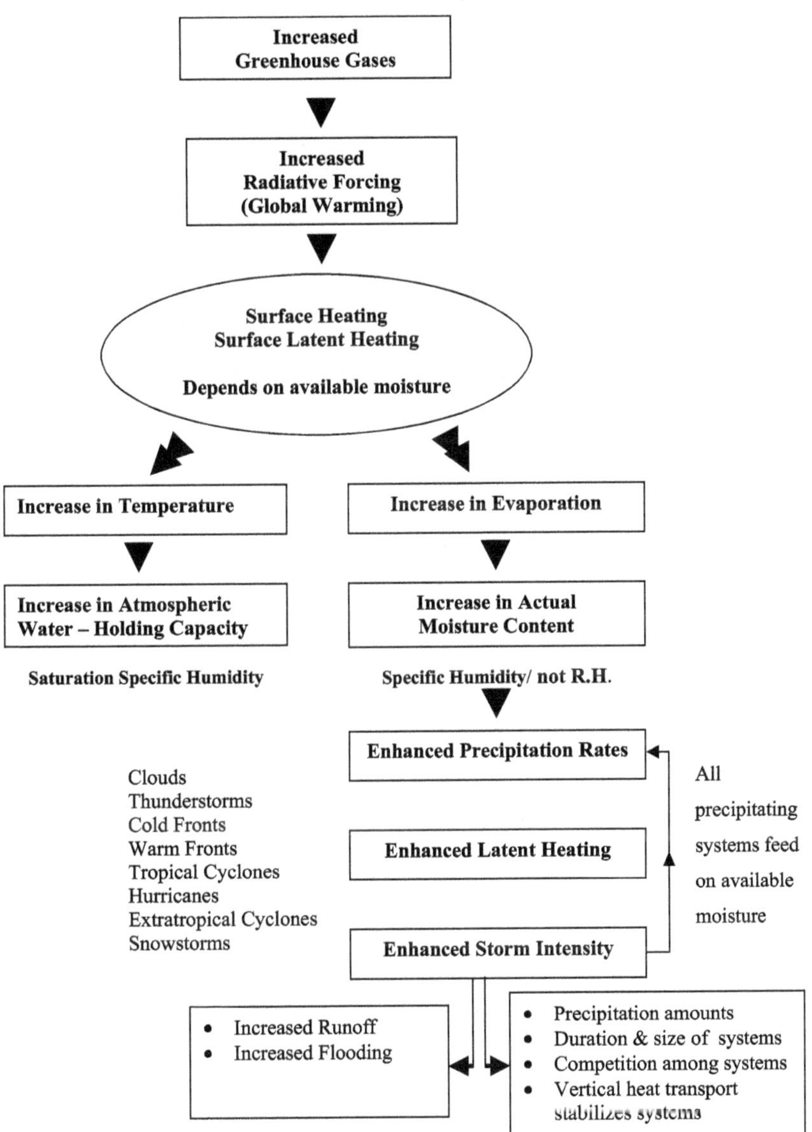

Fig. 10.9 Intensified hydrological cycle.

Source : Trenberth and Shea, (1996).

Existing data indicates that the climate is becoming more extreme in some areas. Some variables which can modulate the global hydrologic cycle might be intensifying in response to global temperature increases. This could translate into prospects for more severe droughts through the depletion of the surface water resources, or floods due to increase in surface water flows. Thus, increase in GHGs not only results in higher surface temperatures (through an increase in atmospheric down welling of infrared radiation primarily due to increased water vapour), but also may strengthen the hydrological cycle through enhanced evaporation and greater water holding capacity of the atmosphere.

Apart from the climate change due to global warming the prevalence and the modulated quasipermant systems of the world under the changed scenario govern and teleconnect the performance of the circulation patterns of major rain bearing components of the weather systems of the world - the Indian Ocean monsoons are a case in point in the context of ENSO and LNSO signals.

The hydroclimatic anomalies in India in the context of perturbation in the regional hydrological cycle, is briefly discussed here. The annual precipitation in space and time over India is about 4,000 km^3, out of which the seasonal rainfall contribution is 3,000 km^3. The annual natural flow in India is about 1880 km^3 , but the availability of water is limited to 1,100 km^3 due to topographic and hydrologic constraints (Anonymous, 1988). The quantum of basinwise annual utilisable water resources is subjected to primarily interannual and intraseasonal variations with respect to the performance of the weather and monsoon systems pattern over India that are modulated under intensified global hydrological cycle through LNSO and ENSO signals.

All India basic water budget elements through the revised water balance procedure during LNSO and ENSO years along with the normal are presented in Table. 10.4. The normal hydrologic regime of all India is dry, subhumid. All India's hydrological regime is improving during LaNina periods. The year 1917 witnessed an increased water surplus of three and half times the normal and was an extreme wet year that coincided with the LaNina phenomena and might have contributed for the betterment of All India's hydrologic regime from dry subhumid to moist subhumid. The years 1965,1972 and 1987 imparted to All India impoverished hydrologic regime by depriving it of its normal dry subhumid status and even pushing it to semi arid status (D) characterised by subnormal seasonal rains. This is reflected in low flows of Indian river systems, as all India recorded zero water surplus from a normal water surplus of 86 mm.

Table 10.4 All India basic water budget elements (mm) - Extreme events

Category of Epoch/Year	Annual Rainfall	Annual PE	Annual WD	Annual WS	Climate Type
Normal	1088	1432	430	86	-24(C_1)
1901-1930	1066	1419	415	60	-25(C_1)
1931-1960	1129	1432	422	117	-21.3(C_1)
1961-1990	1073	1438	442	83	-25(C_1)
Extreme Wet (1917)	1349	1346	289	306	1(C_2)
Good Monsoon (1975)	1201	1400	421	205	-15(C_1)
LaNina (1988)	1160	1467	498	199	-20(C_1)
Extreme Dry (1905)	923	1393	479	14	-33(C_1)
ElNino (1965)	872	1436	544	0	-37(D)
ElNino (1972)	850	1435	536	0	-37(D)
ElNino (1982)	980	1439	464	15	-31(C_1)
ElNino (1987)	967	1475	516	0	-34(D)

The year 2002 was an ENSO year all over the world and its impact was severe with reference to performance of Southwest monsoon over India that recorded a deficit rainfall of 19%. The year 2003 also witnessed a poor performance of southwest monsoon especially over Krishna river basin that forced the farmers to switch to low water-need crops. Fig. 10.10 (source: (Sarma, Padma Kumari & Srinivas, 1999) shows the relationship between the high variability in the flows of Krishna river at Vijayawada point and ENSO/LNSO signal. Generally, high river flows tend to be associated with sub-normal sea surface temperatures and high southern oscillation index, while low river flows are related to abnormal sea surface temperatures and negative southern oscillation index. The

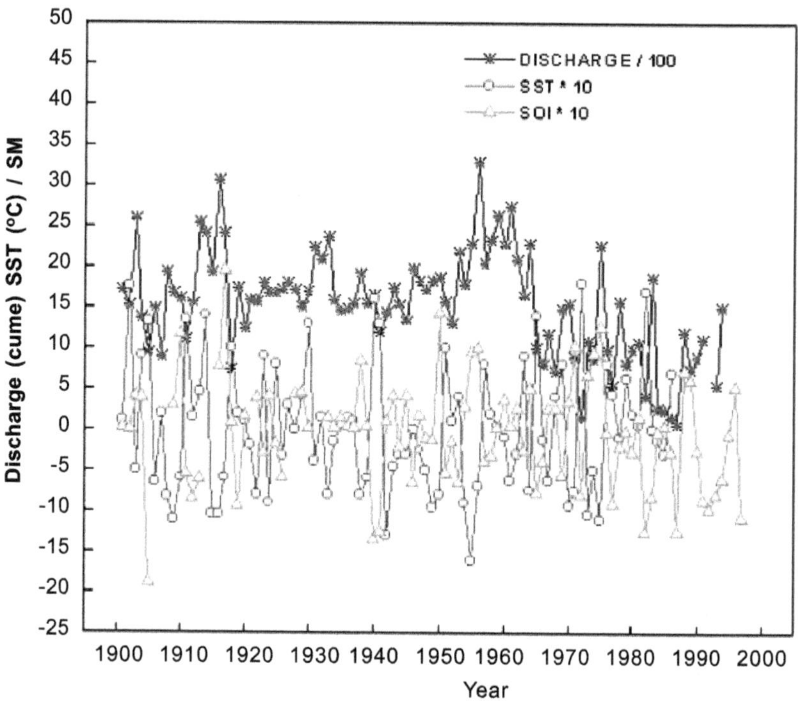

Fig. 10.10 March of annual discharge of Krishna river at Vijayawada point - SST and SOL.

multiple regression analysis of river flows with southern oscillation index and sea surface temperatures of NINO-3 region of Pacific ocean clearly indicate coupled ocean – atmospheric impacts through teleconnected phenomena resulting in either increased or decreased river flows compared to normal during LNSO or ENSO events (Sarma, Padma Kumari & Srinivas, 1999).

Present State of World's Water Wealth

According to Hall and Shiklomanov (1998), the average value of renewable water resources of the world is 42,600 km^3 per year. The largest water resources are in Asia (13,500km^3) and S.America (12,000 km^3). The lowest absolute values of water resources pertain to Europe (2900 km^3) and Australia/Oceania (2,400 km^3). The current world estimate of the total withdrawal is about 3,750 km^3 per year with consumption about 2280 km^3 per year (i.e., 61% of the total water withdrawal). The total withdrawal has been growing by about 10 – 12 % for the past decade and at this rate will reach 5,200 km^3 per year by 2025 (i.e., 1.38 fold increase). But water consumption is increasing at a slightly slower rate (i.e., 1.26 fold increase). Presently about 57% of the total water withdrawal with a consumption of about 70% of the global water is in Asia where irrigated lands are mainly located. Increased human activity and the effects of climate factors on water resources tend to get accentuated in dry climates under changed climate scenario. The climate features determine not only natural values of runoff but to a large extent the rate of change in river flows.

The research activities of the world climate research program (WCRP) and Global Energy and Water cycle Experiment (GEWEX) include assembling available surface measurements and integrating them with meteorological data and satellite observations to produce climatological records of important parameters such as precipitation, river runoff, cloud characteristics, surface radiation and atmospheric humidity. The research activities organized by GEWEX are making significant progress toward the goal of understanding and predicting variations in the global hydrological cycle. GEWEX is making crucial contributions in improving our understanding of seasonal to interannual variations of climate system with reference to global hydrological regime, through its research in hydrological processes, radiation flux and developing data sets for the International Climate Research Community.

The global agricultural, industrial and domestic use of water has increased greatly (Fig. 10.11, source: Houghton, 1997). If water resource managers aim to plan for a sustainable supply for the ever-increasing populations, it is essential to strike a balance between the urban users and the agricultural users who grow food to feed the urban populations. The use of irrigation water for agricultural purposes has increased greatly in efficiency in most developed countries throughout the world.

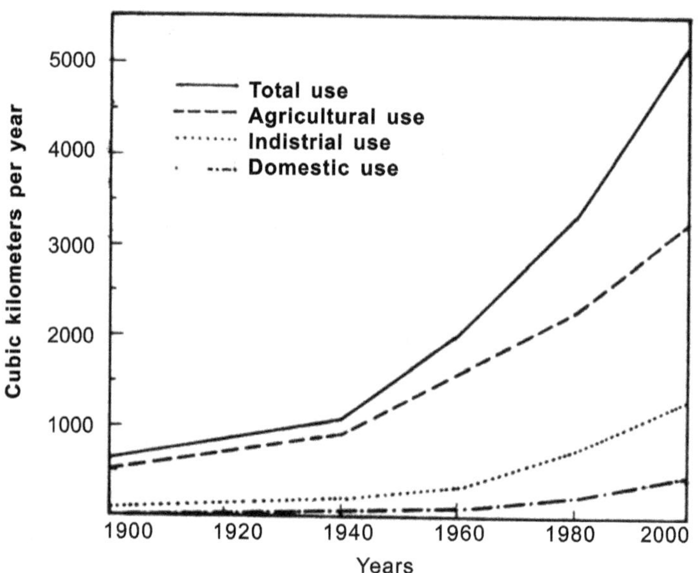

Fig. 10.11 Global water use.
Source : Global warming - The complete briefing.
Houghton (1997).

Even though the world population growth was roughly parallel with a rise in the water use over the past century, the past quarter century has shown a tendency in leveling off the water use despite continued population growth (Fig. 10.12, source: Gleick, 1998). This was achieved by adapting water conservation technologies which, however, are not likely to be sustainable indefinitely. Curtailment of water allocation for irrigation in favour of the needs of water for domestic and industrial use, is hence unavoidable.

From a water resource management point of view the uncertainty associated with aberrated climate system will create great challenges for decision makers and operators of water resources systems.

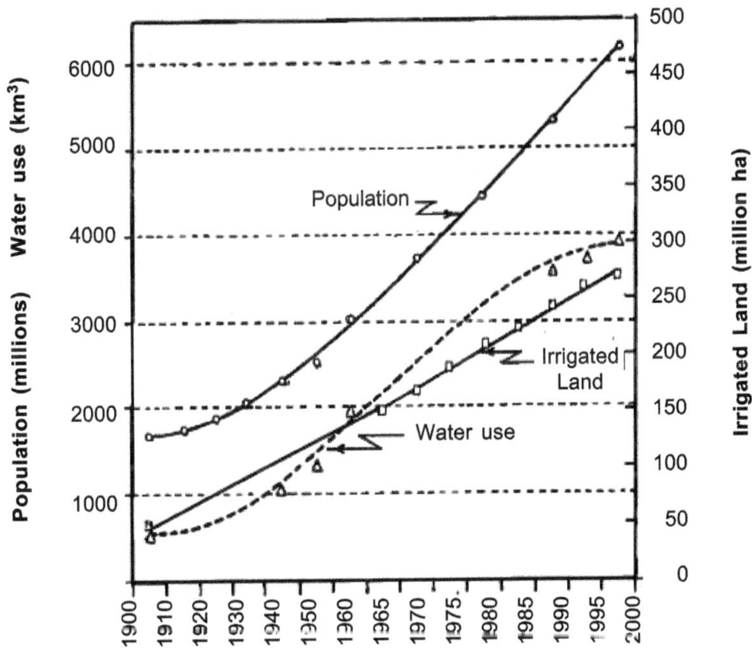

Fig. 10.12 Global population - water use and irrigated land. Source : Gleick (1998) The world's water.

References

Anonymous (1988) *Water resources of India,* CWC Publ no. 30/88, New Delhi.

Gleick,P.H. (1998) *The World's water : 1998 – 1999.* The Biennial report of fresh water resources. Island Press, Washington, D.C.

Hall, A.J., and Shiklomanov, I.A (1998) *Water issues – The climate setting.* WMO/TD-no 904, 78-98.

Houghton, J (1997) *Global warming – The complete briefing.* Cambridge University Press.

IPCC (1996) Inter Governmental Panel on Climate Change (IPCC). Contribution of Working Group I to the Second Assessment Report of IPCC, Meira Filho, L.C, Callander, B.A, Harris, N, Kattenberg,A., and Maskell, K, (eds), Cambridge University Press.

IPCC (2001) *Climate Change, Synthesis report*, IPCC 2001, Cambridge, University Press.

Karl,T.R and Knight,R.W. (1997) The 1995 Chicago heat wave: How likely is a recurrence. *Bull. Amer. Meteor. Soc*; 78, 1107 – 1119.

Kutzbach,J.E and Guetter,P.J (1984) Sensitivity of late glacial and Holocene climates to the combined effects of orbital parameter changes and lower boundary condition changes: Snap shots simulation with a general circulation model for 18, 9 and 6 kaB.P. *Annals of glaciology*, 5, 85-87.

Lambert, S.J (1996) Intense extratropical northern hemisphere winter cyclone events: 1899-1991, *J.Geophys.Res*; .101, 21319 – 21325.

Legates, D.R and Mather,J.R.(1992) An evaluation of the average annual global water balance. *Geogr. Rev.*, 82 (3),253-267.

Nicklas,G.P and Imbrie,J (1986/87) Orbital geometry, CO_2 and Pleistocene climate. *Oceanus*, 29(4), 43-49.

Peixito, J.P; and Oort,A.H (1993) *Physics of Climate*, American Institute of Physics, New York.

Sarma, A.A.L.N, Padma Kumari, B and Srinivas.S (1999) Studies on hydrological extremes - ENSO signal. IAHS Publ 255, "*Hydrological Extremes: Understanding, Predicting, Mitigating*", p. 73-80.

Thornthwaite, C.W. & Mather, J.R. (1955) The Water Balance. *Publ. in Climatology*, v.8, 100- 104, Drexel Inst. Technology, N.J.

Trenberth,K.E and Shea,D.J (1996) Atmospheric circulation changes and links to changes in rainfall and drought. 7th Conf.Climate variations, Amer .Met. Soc, 14-19.

11

Impact of Land-use and Land-cover on Water Quality

T. Vijaya Lakshmi

Centre for Environment, Institute of Science and Technology
Jawaharlal Nehru Technological University, Kukatpally, Hyderabad-500 072
E-mail : tatiparti@yahoo.com

Key words : Land Use; Land cover; Remote Sensing, GIS, Water quality Index; Nakkavagu case history; Remediation measures

Introduction

Though industrialization is essential for the economic development of a country, it is not an unmixed blessing. All industries generate wastes, and in the ultimate analysis, all wastes have an adverse impact on water (surface or groundwater), sooner or later. Leachates from solid wastes (say, from mineral industries) and toxic aerosols (say, from thermal power plants) reach the groundwater via the soil cover Toxic effluents from (say) chemical and pharmaceutical industries reach the surface or groundwater faster and in greater proportion, and hence their adverse impact on water is most serious. It is for this reason that the environmental protection regulations specify stringent limits for compliant effluents.

The environmental degradation which is evident in the catchment area of Nakkavagu stream in Medak District, A.P., India, shows how an environmental disaster could occur if industrialisation takes place without regard to the protection of the environment. It is tragically ironical that this currently degraded wasteland was once known as methukudurgam symbolizing the abundance of paddy grown in the area.

Biophysical Setting of the Nakkavagu Catchment Area

The Nakkavagu catchment area (711 km^2 ; long. 78^0 05' and 78^0 25' E. and the lat. 17^0 25' and 17^0 45' N) constitutes 7.5% of total geographical area of the Medak distract..

Patancheru and Bollarum are the most important industrial areas in the Nakkavagu catchment (Fig. 11.1).

The district forms part of the Deccan Plateau and major part of the district falls under Godavari basin. Medak district is a backward district with more than 80% of the population dependent on agriculture. It has low literacy rate and low per capita income. The predominant soils in the district are black and red soils. The climate of the Medak district is characterized by hot summer during the period from March to May followed by the south west monsoon from June to September. October and November forms the post monsoon season and cold season commences from December and continues up to early February. Precipitation, temperature relative humidity and wind velocity and evapo- transpiration are the important factors that influence the climate of the region. May is the hottest month with mean daily maximum temperature 40^0 C and mean daily minimum temperature 26^0 C .The day temperature often exceeds 42^0 C. During cold season, the night temperature may go down to about 7^0 C. The mean monthly temperatures range approximately from 22^0 C in Dec. to 32^0 C in March. The climate of the district can be classified as tropical savannah climate.

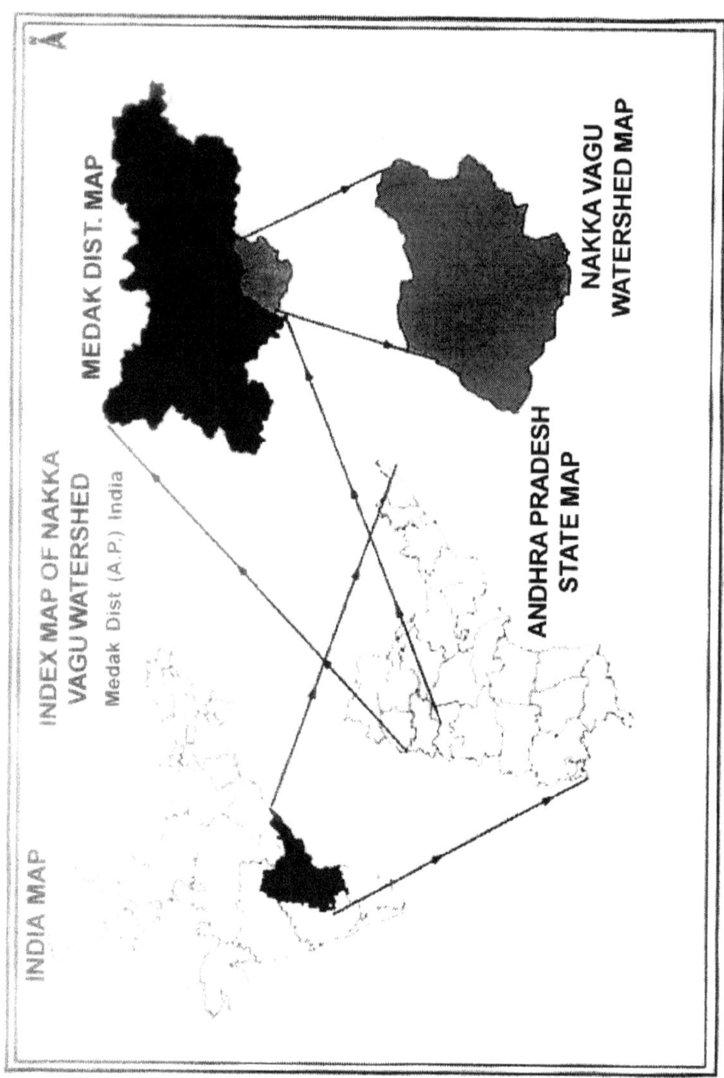

Fig 11.1 Location map-Nakkavagu watershed, Medal Dt., (A.P) India.

On an average there are 54 rainy days in a year in the district and the number of rainy days varies from 52 at Sanga reddy to 57 at Medak. The Medak district receives normal rainfall of 884.5 mm. The rainfall increases from south to north of the district. The average annual rainfall for the 10 years period is 802.8 mm. About 85% of the rainfall is received during south-west monsoon season and the maximum number of rainy days occurs in the month of July. Evaporation increases from February through June. A steep fall in evaporation rate is observed during the first half of south-west monsoon season. Maximum evaporation rate is observed in May at 167.2 mm with minimum occurring in November at 7.6 mm (District Handbook, 1999).

Growth of Industries in the Area

Between 1975 and 1995, within a radius of 30 km around Hyderabad city, a number of drug and pharmaceutical industrial estates (IEs) came up at Patancheru, Ramachandrapuram, Pashmylaram, Yelumela, Zaheerabad, Bollarum, Bonthapally, Kazipally and Gaddapotharam. These industrial estates were located in the Godavari basin of the Nakkavagu stream. The largest IE is at Patancheru, which was developed in an area of 440 ha in five phases. Some of the State's biggest bulk drugs and pharmaceutical industries are located in the IEs at Patancheru and Bollarum. The industrial estates include one ancillary industrial estate for BHEL (Bharat Heavy Electricals Limited) at Ramachandrapuram. Though the industrial development took place to take advantage of the infrastructural facilities organized by the government to promote industrialization of a backward area, it has resulted in deterioration of the environmental quality of the region (Sai Bhaskar Reddy, 1999). During the period, 1992 –1997, the number of large & medium scale industries registered in the district went up from 176 to 259, and the number of small scale industries rose from 3441 to 4056 units.

The concentrated and uncontrolled development of the industries, particularly drug industries, had a strong adverse impact on the ecosystem and water quality of Nakkavagu drain. In view of these, APPCB has banned setting up of new industries in nine industrial areas located in this district for the period of two years.

Industrial Areas

Patancheru Industrial Area

Patancheru developed as the loci of industries, as (a) it is 35 km from the city of Hyderabad, (b) it is on the national highway No: 9, leading to Mumbai, (c) it has infrastructure facilities such as electricity, roads etc and assured supply of water from the Manjeera reservoir, (d) it also lies within the borders of Medak district to take advantage of the incentives and subsidies declared by the government of Andhra Pradesh and Govt. of India for the industrial development of backward districts. Patancheru is the largest industrial estate covering a total area of 400 ha with the maximum density of industries. In this area, there are 122 industrial units of different types (chemical, pharmaceuticals, steel, dyes, paints, engineering etc.). The wastewater from the industrial units is currently discharged into the nearest open drains, leading to mullahs who finally join natural water bodies. A common effluent treatment plant is developed at Patancheru to take care of few of the industrial effluents (Anjaneyulu, 1998)

Bollarum Industrial Area

The Bollarum industrial area is about 28 km away from Hyderabad. It is a privately owned co-operative estate, with about 60 industrial units which are predominantly related to chemical and pharmaceuticals. A common effluent treatment plant (CETP) has been built to serve this industrial area.

Impacts of land Use and Water Pollution

All land uses have an effect on water quality, whether positive or negative. Land use is in part determined by environmental factors such as soil characteristics, climate, topography, and vegetation. In fact, land use and practices are probably the most important factor in determining water quality in most of the watersheds. Land use and water quality are directly linked. Initially, the quality and quantity of water available in an area plays a major role in the choice of use to which a given piece of land is put (crop land, pasture, habitation, industries, recreation, etc.). Per contra, the use to which land has been put would determine the nature and extent to which the water quality will be affected (biological pollution from habitations, chemical pollution from industrial use, pesticide and fertilizer pollution from agricultural use, etc.) (Aswathanarayana, 2001).

Various land uses can potentially introduce many pollutants into aquatic ecosystems. Elements such as Copper, Zinc, Manganese, Boron and Phosphorus, are toxic in high concentrations, though they constitute essential elements (with Recommended Daily Intake) in the case of humans and for the support and maintenance of aquatic ecosystems. The concentrations above which water pollutants adversely affect a particular water use may differ widely. Water quality requirements, expressed as water quality criteria and objectives, are use-specific or are targeted to the protection of the most sensitive water use among a number of existing or planned uses within a catchment.

Remote Sensing and GIS Approaches for Quantification of LU-LC

For the assessment of impact of land use / land cover changes on water quality, spatial and non spatial database has been generated using remote sensing, GIS and field studies (Anjireddy, 2001). Non Spatial database consists of water quality estimation (Standard Methods for examination of water and wastewater, (1998) and calculation of water quality indices (Srivastava and Sinha; 1994). Spatial database consists of mapping of land use / land cover of Nakkavagu catchment for the years 1971, 1981 and 2001. Methodology adopted for generation of LULC is given in Fig. 11.2. Survey of India topographical maps were used to prepare the land use / land cover map of 1971.

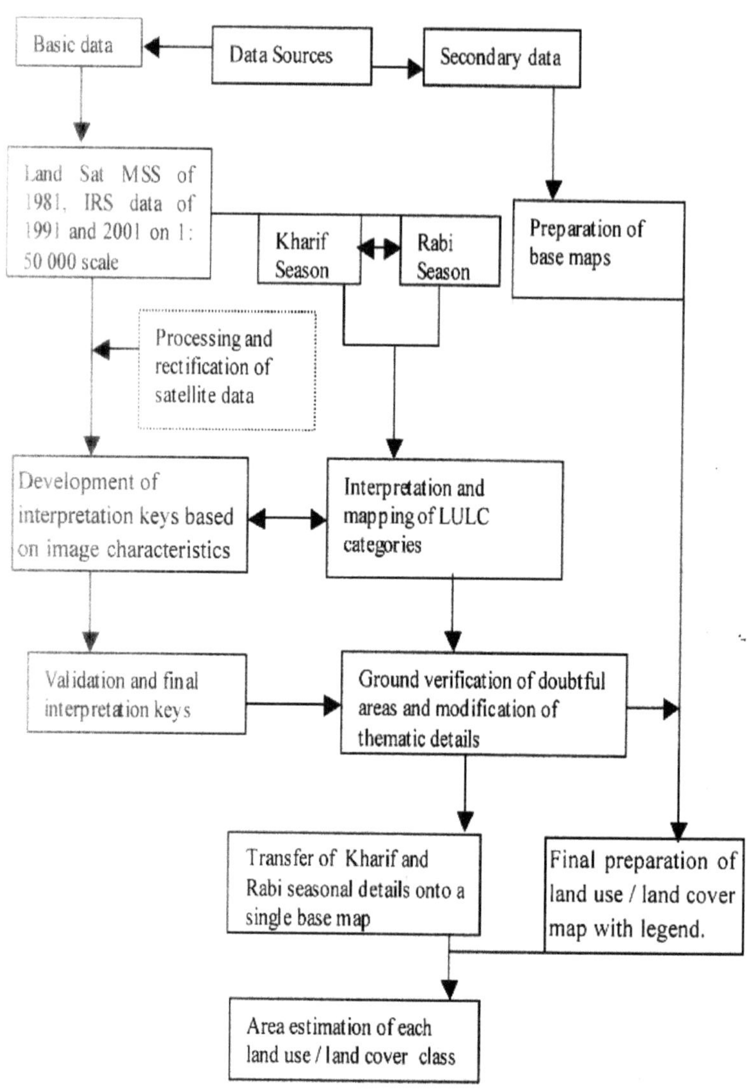

Fig 11.2 Flow chart showing methodology for land use / land cover mapping

Remotely sensed satellite data for the years 1981 and 2001, was acquired from National Remote Sensing Agency, Balanagar, in digital form. MSS of Land Sat, date of pass May 8th 1981 and LISS III of IRS ill, date of pass of 20th August 2001 are obtained. The digital data is then processed, rectified, and enhanced for delineation of study area. From the full scene, the study area Nakakvagu watershed is extracted by clipping the data with the drainage boundary in EASI / PACE 6.3 ver. The data is then composed in ACE environment, for taking hard copy of false color composite of the satellite data for further interpretation and preparation of land use / land cover. Using the rectified, enhanced digital satellite data of 1981 and 2001, land use / land cover map is generated using digital classification technique by supervised classification.

The imagery is interpreted and ground checked for corrections. The final map is prepared after field check. Classification of different land use / land cover categories in the study area has been carried out based on the NRSA land use / land cover classification system (Table 11.1).

Table 11.1 Land Use / Land Cover Classification System in Nakkalavagu

Level-I	Level-II	Level -III
1. Built up land	Towns / Cities Villages	
2. Agricultural land	2.1 Crop land + 2.2 Fallow	
3. Forest	3.1 Deciduous (moist and dry) 3.3 Scrub forest	3.1.1 Open forest 3.1.2 Dense Forest
4. Waste Lands	Barren rocky / stony waste-quarry / sheet rock area	
5. Water Bodies	5.1 River / Stream 5.2 Lake / Reservoir / Tanks	

The digital GIS database is created in order to derive the change detection map showing the variation of land use / land cover, and the maps showing the spatial distribution of water quality index.

Major land use / land cover change in Nakkavagu catchment area is in the form of built-up land and the change accounts for 165.08 % of the total geographical area from 1971 to 2001. And these changes are mainly due to the proximity to the Hyderabad city, and a National Highway-9 passing through. With the help of field survey, it is observed that most of the land changed to built-up class is mainly under industrial developmental areas (IDA's). In Patancheru and Bollarum most of the crop land has been changed to built-up land. In 1971, in the total geographical area, forests occupied 4.6 % , streams 0.5 % , waste lands 22.4 %, tanks 5.6 % , crop land is 63.8 % and built-up land 3.1 %. Fallow lands were not identified from the Survey of India toposheet of the year 1971 (Table 11.2, and Fig. 11.3).

Land use land cover changes from 1971-1981-2001 are studied by overlay analysis in ARC/ INFO GIS platform, using land use / land cover maps of 1971, 1981 and 2001. The change detection maps for the years 1971-81, 1981-2001 and 1971-200 I are presented in Fig. 11.4.

Table 11.2. Land use / Land cover Statistics of Nakakvagu Catchment Area

(Area in hectares)

	LULC-1970	LU-LC-1980	LULC-2001
Forest	3305.4375	3204.375	3203.1875
Stream	334.375	323.25	286.375
Waste land	15927.75	15051.125	9473.375
Tanks	4019.0625	4000.4375	3559.25
Crop land	45406.1875	41114.0625	26240.3125
Fallow	0	4025.625	22618.4375
Built-up land	2187.875	3461.75	5799.6875

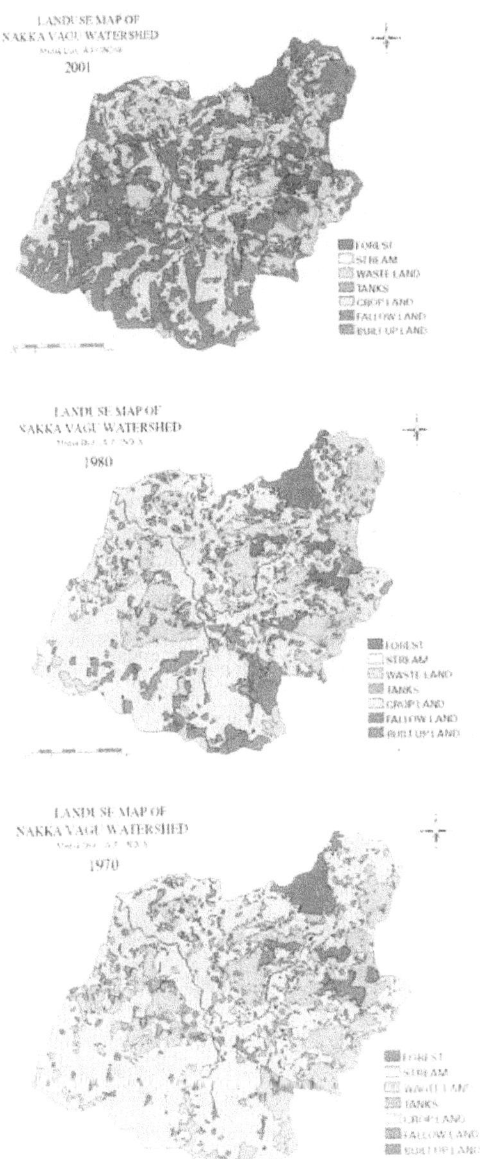
LANDUSE MAP OF
NAKKA VAGU WATERSHED
2001
FOREST
STREAM
WASTE LAND
TANKS
CROP LAND
FALLOW LAND
BUILT UP LAND
LANDUSE MAP OF
NAKKA VAGU WATERSHED
1980
LANDUSE MAP OF
NAKKA VAGU WATERSHED
1970

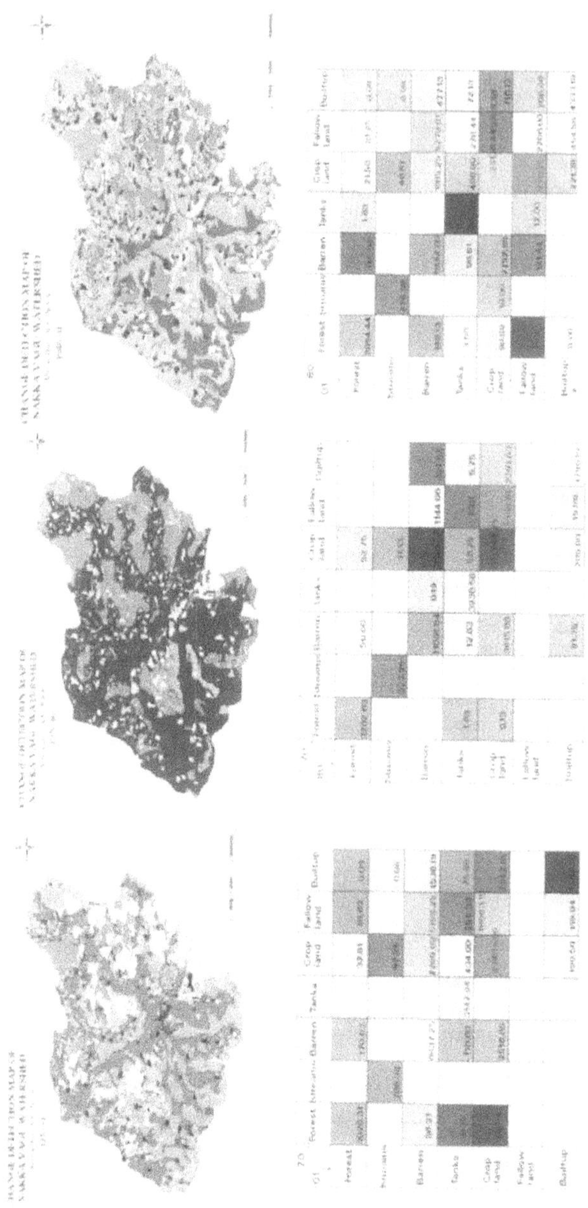

During 1971-80 there was a decrease of forest area by 3.05 % while between 1981-2001 there was a slight change in the forest area amounting to 0.03 %. It is observed that the forest cover decreased by 3.09 % during 1971- 2001. This is indicative of the realization of adverse impact of deforestation which was curbed by the rigorous enforcement of laws. In fact protective forestation principles are coming in to focus during this period. But the area under streams decreased between 1971-80 by 3.33%, and between 1981-2001 stream area decreased by 11.40%. And over a period of 30 years i.e., 1971-2001 stream area decreased by 14.35%. This decrease in stream area is the symptoms of urbanization and industrialization due to increase in population. Wastelands also decreased by 5.50% in the watershed area during 1971-80, between 1981-2001 the change in waste lands was 37.05% and during 1971-2001 the change observed was 40.52%. The wastelands have been converted into built up land between 1981-2001.

In the Nakkavagu watershed area, good watershed management practices were implemented during 1981-2001. Wastelands have been converted into cropland because of the availability of dug well / borewell irrigation However, due to the extensive exploitation of groundwater resources, groundwater contamination also increased drastically. In the Nakakvagu watershed area, during 1971, tanks were occupying 5.6 % of the total geographical area. This area is decreased to 5.0 % by 2001. There was no change in the tank area between 1981-2001. When tank area is considered as a whole there was a decrease of 11.44 % of the area between 1971-2001, and it is 11.02% between 1981-2001, with a very negligible change of 0.46% during 1971-80. This shows that tank area over the years has been converted into cropland, fallow land, and built up lands, leading to the pollution of the surface water, groundwater and soil, and solid waste dumping resulted in the loss of soil fertility. There is a decrease of 42.20% of the cropland between 1971-2001, with a decrease of 36.17% between 1981-2001 and 9.45% between 1971-80.

This LV / LC change pattern clearly indicates how one kind of land use affects the land cover. Thus there was an increase of 6.2 % of fallow land between 1981-2001, which was mainly due to the conversion of 60 % of the cropland to fallow land. The change in built-up land was 58.21 % during 1971-80, 67.5 % during 1981-2001, and 165.08 % during 1971-2001. Major land uses contributing to the increase in built up land between 1971-2001 was wastelands (26.48%), tanks (1.3 %), crop land (40 %), and fallow land (3.80 %). This critical examination of land use / land cover change detection analysis leads to the conclusion that the conversion of one type of land use, if done without proper. Perspective, leads to overall environmental deterioration and adversely affects sustainable development.

Spatial Distribution of Water Quality Index

Spatial distribution of water quality index maps has been generated using GIS ARC INFO software. Earlier investigations revealed that the Nakkavagu catchment is facing severe environmental problems due to water pollution. The Nakkavagu and its tributaries namely, Pamulavagu, and Iskavagu are suffering from the contamination caused by domestic sewage, industrial effluents, and agricultural non point source pollution. The Supreme court directed NEERI (1994) and CPCB (1998) to carry out a study for pollution evaluation and status of these streams with respect to environmental quality parameters and problems. Keeping this in view, a number of studies were conducted and reports were submitted to the Government for remediation of this catchment. But there is no clear cut mandatory for restoration of this situation. Hence an attempt is made to create a mandatory buffer of these suffered streams with the help of line buffer techniques. The line buffers along Nakkavagu and other streams are created by providing 5 km on either side of the stream as an area of influence. 5 km buffer is considered because of the following reasons :

(i) The human habitats / the major industries are located within the radius of 5 km and beyond as per the land use / land cover map.

(ii) As the terrain slope towards the channels is steep, the flow conditions are favorable to carry the effluents from uplands to the stream course.

(iii) The groundwater quality is below the permissible limits for potable water up to 5 km in most of the areas and beyond at few places. This is evident by the information derived from hydro geomorphology map.

Significance of Buffers

The setting of specific buffer zones or separation distances has the obvious advantage of providing a clear indication to a developer of siting preferences and limitations which should minimize the potential for disagreements and provide considerable assistance to decision makers in determining applications. The surface could be visualized as representing a third dimension to the 2-Dimensional x-y data and the third axis can be represented by any attribute. Spatial distribution maps for water quality are prepared based on the index computed from the results of the laboratory analysis using ARC INFO software.

Spatial Distribution of Water Quality

Spatial distribution of water quality index maps are generated for 6 years namely 1979, 1984, 1989, 1994,1999 and 2001. The ranges of water quality indices for 5 point scale are as follows:

0-25	:	Excellent
26-50	:	Good
51-75	:	Poor
76-100	:	Very poor
> 100	:	Unfit for drinking

The entire study area consists of a number of polygons, and all these polygons are categorized into five groups for preparation for spatial distribution of water quality index and each group is colored with a different color which is shown in the maps. Water quality data is collected from earlier reports (A.P. State Ground water board, District ground water department, NGO organizations, CPCB reports, NGRI, IPM) for the years 1979, 1984, 1989,1994, and monitored from 1999-2001 for the villages namely Mutangi, Ismailkhan pet, Arutla, Gaudacherla, Patancheru, Nakakvagu at near national highway 9, Palapnur. The villages are selected mainly because of their location in the watershed area. These villages lie near to the Nakkavagu stream and industrial pollution is more evident as the streams and tanks carry spillovers / treated / untreated industrial effluents and domestic waste. And more so, water quality data of these villages is available from 1979 onwards and for comparative study with the present monitoring locations.

Estimated water quality index taken as z value for the generation of spatial distribution maps of water quality, as a function of 3 dimensional surface modeling. The computations of quality indices for water are shown in Table 11.3. According to earlier investigations carried out by several authors, a five point scale, namely excellent, good, poor, very poor, and unfit for drinking for water quality has been selected.

Table 11.3 Deterioration in water quality index in some villages of Nakkavagu catchment area as observed during 1970's, 80's, 90's and 2001.

S.No		Year	WQI	Rating
1	Muttangi	1979	20	Excellent
		1984	41	Good
		1989	59	Poor
		1994	76	Very Poor
		1999	94	Very Poor
		2001	96	Very Poor

Table 11.3 *Contd...*

S.No		Year	WQI	Rating
2	Ismailkhanpet	1979	21	Excellent
		1984	39	Good
		1989	50	Good
		1994	63	Poor
		1999	78	Very Poor
		2001	99	Very Poor
3	Arutla	1979	22	Excellent
		1984	46	Good
		1989	64	Poor
		1994	77	Very Poor
		1999	63	Poor
		2001	90	Very Poor
4	Gaudacherla	1979	19	Excellent
		1984	45	Good
		1989	49	Good
		1994	50	Good
		1999	69	Poor
		2001	91	Very Poor
5	Patancheru	1979	-	NA
		1984	-	NA
		1989	63	Poor
		1994	82	Very Poor
		1999	103	Unfitfor Drinking
		2001	101	Unfit for Drinking

Table 11.3 *Contd...*

S.No		Year	WQI	Rating
6	Nakkavagu	1979	41	Good
		1984	42	Good
		1989	73	Poor
		1994	75	Very Poor
		1999	107	Unfit for Drinking
		2001	110	Unfit for Drinking
7	Palapnur	1979	11	Excellent
		1984	18	Excellent
		1989	42	Good
		1994	48	Good
		1999	63	Poor
		2001	65	Poor
8.	Khazipally Tank	1999	81	Very Poor
9.	Gandigudem	1999	96	Very Poor
10	Isnapur Village	2002	70	Poor
11	Pocharam	1999	94	Very Poor
12	Ganapatigudem	1999	98	Very Poor
13	Bachugudem	1999	90	Very Poor
14	Patancheru	2000	147	Unfit for Drinking
15	Lakdaram	1999	74	Poor
16	Peddakanjeral	1999	75	Poor
17	Chedruppa	2001	90	Very Poor
18	Byathole	1999	70	Poor
19	Kistareddy pet	2000	82	Very Poor
20	Khazipally Village	1999	103	Unfit for Drinking

On the critical examination of the spatial distribution maps of water quality indices, the following conclusions are drawn.

Spatial distribution map of water quality index is given in Fig. 11.5. During the year 1979, water quality index ranged between excellent and good. Most of the area was under crop land and agricultural practices are to the maximum. Because of this, agricultural non-point source pollution is mainly contributing to the quality. More so the domestic water pollution. Spatial distribution map is well correlating with the land use practices.

In 1984, the spatial distribution map of water quality index map, indicates excellent and good quality indices are found in this region. The areas namely Gaudacherla, Arutla, Chedruppa, Ismailkhan pet, Biathole, and Isnapur have water quality in excellent category in 1979 are converted to good water quality category in 1984. The change in water quality is mainly due to discharge of industrial effluents.

By the year 1989, pollution has increased and is clearly reflected in the water both surface and ground water quality, and soil contamination and is evident from the spatial distribution map of water quality index. Because of the increased usage of fertilizers, pesticides non point source pollution from agricultural activities also increased apart from the industrial pollution. Almost all the areas are under poor water quality category except Gaudacherla and Palapnur. Water quality during 1994, is almost in all the areas is under poor and very poor classes. Many of the researchers, scientists have submitted their evaluation reports to A.P. Pollution Control Board, High Court, and to Supreme Court. Areas namely Khazipally tank, Khazipally village, Kistareddy pet, Patancheru, Pocharam, Muttangi, Nakkavagu, Ganapatiguda, Gandigudem, have reached very poor water quality. And in villages Arutla, Bachugudem, Byathole, Chedruppa, Gaudacherla, Ismailkhanpet, Isnapur, Lakdaram, Peddakanjerla (Kanjerla Bhadruk) water quality index became poor by the year 1994. The results were tallied with the field observations like improper functioning of effluent treatment plants, the increase in number of registered industries in these industrial areas, and disposal of solid and hazardous waste etc. By the year 1999, because of the increased exploitation of land like sand mining, quarrying, improper utilization of land etc., water quality in areas Patancheru, Nakkavagu and Muttangi has become unfit for drinking. And in the areas Pocharam, Ganapatiguda, Isnapur, Bachuguda, Peddakanjerla (Kanjerla Bhadruk), Ismailkhanpet is very poor and in the areas Lakdaram, Biatul, Chedruppa, Arutla, Gaudacherla, and Palapnur water quality is rated as poor. From the spatial distribution of water quality map of the year 2001 it is seen that except Palapnur all the areas are classified under very poor and unfit for drinking. It is observed that from the spatial distribution maps after 1984 quality of water resources started deteriorating.

Conclusions

According to press reports (*The Hindu* of Nov. 5, 2004), pollution in Patancheru area has reached catastrophic levels -nine tanks and streams in Medak district are unusable, and another three are on the verge of becoming so. Out of the 22 ponds studied by the fact-finding committee of the A.P. High Court, only four ponds are free from toxic chemicals. The Committee recommended immediate remediation of Khazipalle and Asanikunta tanks, as any further delay would choke downstream tanks of Kistareddypet and Gandigudem. The Committee recommended that the construction of a pipeline to carry the treated industrial effluents from Patancheru to the Common Effluent Treatment Plant (CETP) at Balanagar be expedited. The Environmental Protection and Training Institute (EPTRI) found that 30 % of the soils have high chloride and sodium contents which have the effect of greatly reducing the agricultural productivity of the soil. It is unclear as to what extent the soil pollution is attributable to industries and whether it is partly due to overuse of soil.

It is evident that the land use and land cover changes in the form of converting crop land into built-up land (for habitations and industries) and fallow land, and decrease in forest land, had a strong adverse effect on water quality and soil fertility. While it is virtually impossible to restore *status quo ante* in the case of Nakkavagu watershed, it is possible to mitigate the situation. Any strategy for the amelioration of the situation, should have the following components: (i) Groundwater: When once the aquifers are heavily polluted, as is apparently the case in Nakkalavagu, it is technically difficult and prohibitively expensive to detoxify them. Heavily Polluted groundwater can only be cleansed of the pollutants only at the point of abstraction. If, however, the pollution of groundwater is slight, amelioration can be attempted through the blocking of the sources of pollution, and allowing natural processes of regeneration a chance to ameliorate the situation, (ii) Surface water and soils: These can be ameliorated through a combination of liming, and bioremediation, phytoremediation and wetland treatment technologies. Customised microbial systems are available to ameliorate any kind of polluted water through constructed wetlands (Brierley, 1990).

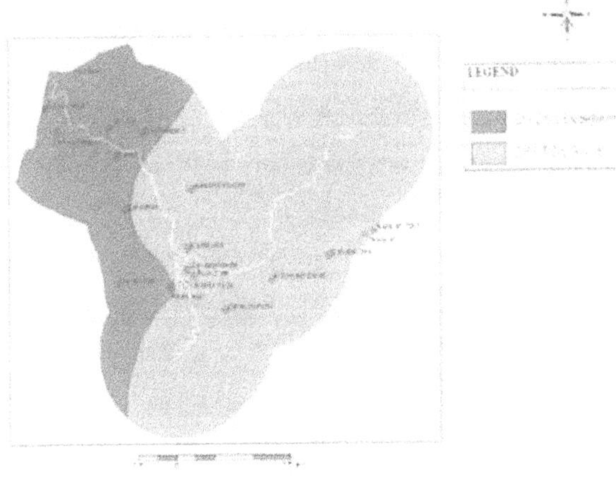

(a)

SPATIAL DISTRIBUTION OF WATER QUALITY INDEX 1989
PART OF NAKKA VAGU WATERSHED
Medak Dist. (A.P) INDIA

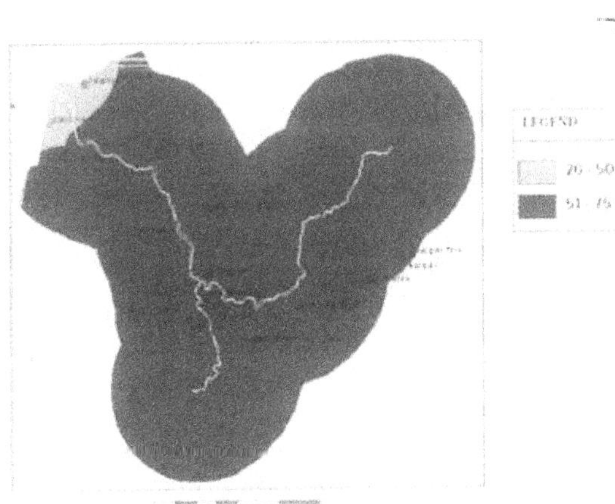

(b)

SPATIAL DISTRIBUTION OF WATER QUALITY INDEX - 1999
PART OF NAKKA VAGU WATERSHED
Medak Dist. (A.P) INDIA

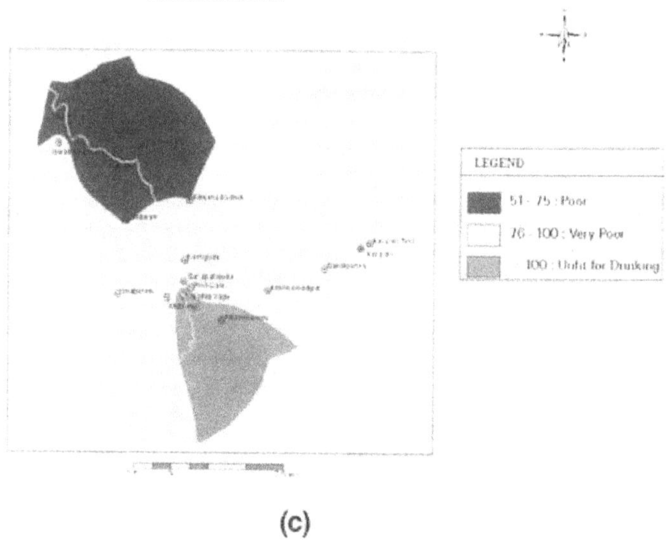

(c)

References

Anjaneyulu Y, (1998). Final Technical Report on Conservation and Management plans for Khazipally cheruvu, submitted to APPCB.

Anji Reddy M. (2001). *Text book of Remote Sensing and Geographical Information System,* B.S Publications, Hyderabad.

Aswathanarayana U., (2001). *Water Resources Management and the Environment* Rotterdam: A.A. Balkema.

Brierley C.L., (1990) Bioremediation of metal-contaminated surface and groundwater. *Geomicrobiology, J,* **8**, 202-233.

District Hand Book, Medak District, 1999.

Sai Bhaskar Reddy, (1999). *EIA studies on Patancheru and Bollarum areas,* Unpublished thesis submitted to Osmania University

Srivasta A.K; Sinha D.K (1994). Water quality index for river Sai at Rai Bareli for the pre-monsoon period and after onset of Monsoon. *Ind. .J of Environ Protection,* **14**, No 5, May.

Standard methods for examination of Water and Wastewater (1998). American Public Health Association, Washington, D.C.

12

Geomatics for Watershed Management

M. Anji Reddy

Centre for Environment, Institute of Science and Technology

Jawaharlal Nehru Technological University, Kukatpally, Hyderabad-500 072

E-mail : mareddyanjireddy@rediffmail.com

Key words : Watershed as a system ; Remote Sensing ; GIS, Data sources; Integrated Watershed Management ; Benefits.

Watershed management involves an integrated and inter-disciplinary approach. It generally requires landuse adjustment measures, which contribute to the reduction in soil erosion rates, increased agricultural production, generation of rural employment and balanced growth of national economy. By adopting the watershed as a unit, different measures are adopted and executed carefully in each of the toposequences according to their capability. It essentially relates to soil and water conservation in the watershed which includes proper land use, protection of land against all forms of degradation, building and maintaining soil fertility, water conservation and harvesting, proper management of surface and ground water, flood protection, reduction in the quantity of sediment, increase in productivity from all land uses etc.

Integrated watershed management has come to be recognized internationally as an important holistic approach to natural resources management, which seeks to promote the concept of sustainable development. The effort requires a systematic and strategic integration of water resources development and management activities into the social and economic development process. Sustainable land use forms an overall planning framework, whilst sound land-use planning concepts, together with the adoption of appropriate land-use practices, provide key guidelines for land and water resources development and management, which should be undertaken with the integrated objectives of reducing natural disasters, boosting productivity and achieving sustainable development.

Sound land-use planning methods and practices can be developed from an end-use standpoint, such as social and economic development of national or regional planning, or from a sectoral point of view, i.e. in the context of development planning for various sectors such as agriculture, forestry, mining and water resources. Integrated land-use planning aims to address the linkages, especially the links between national and sectoral planning, to form a basis for better planning in the fields of land use and water resources development. The important elements of linkages include the management system, financial resources, institutional and legal frameworks and community participation. The publication also identifies key areas, which help to focus the land-use planning process towards the effective reduction of water-related natural disasters and also towards improved watershed management. Sustainable watershed management is to coordinate land and water resources development activities with the development activities of other sectors, and to adopt an ecosystem-oriented foundation for the prioritization of water-related development activities.

Watershed as a System

Watershed Development and Management (Institutional Inputs)

Evapotranspiration

Precipitation and sunshine (natural inputs)

Technology agro chemicals, technological

The watershed relief soils the population and natural resources use vegetation infra-structrure institution/ services

Runoff

Dam and reservoir

Irrigation electricity recreation other

Products : livestock crop forest others

Population requirements

Undesirable by-products (INSITU)

Erosion (B) others

Downstream
(a) Sedimentation
(b) Others

Causes of Degradation of Watershed

Uncontrolled, unplanned, and unscientific landuse and intervention leads to degradation of watershed area. Some of the contributory activities are:

1. **Agricutural Land**
   - (a) Cultivation on sloping lands without adequate precaution
   - (b) Cultivation without agronomic measure to conserve soils and waters.
   - (c) Cultivation along susceptible nalla banks.
   - (d) Overcropping, faulty agricultural techniques.
   - (e) Cultivation of erosion permitting crops.
2. **Grassland**
   - (a) Excessive and uncontrolled grazing.
   - (b) Growth of weeds.
   - (c) Development of cattle tracks.
   - (d) Fires and thefts.
3. **Forestland**
   - (a) Excessive and uncontrolled grazing.
   - (b) Clear filling on steep slopes.
   - (c) Fires and thefts.
   - (d) Biotic pressure for fuel, fodder.
4. **Shifting Cultivation**
   - (a) Damage the protective and productive vegetation.
   - (b) Damage to the topsoil and inhibits the growth of grasses, shrubs and trees.
5. **Unscientific Mining and Construction Activities**

6. Fires

Intentional or accidental fires result in loss of vegetation, organic matter and microorganisms.

7. Noncooperation of the Community

Objectives of Watershed Management

- Study the characteristics of the concerned area of a watershed, namely, the size, shape, slope, drainage, vegetation, geology, soils, climate and land use using Indian remote sensing satellite data.
- To determine the status of erodability.
- To evaluate various watershed characteristics of soil erosion using satellite data.
- To suggest appropriate operational measures to conserve soil and water to improve the ability of land to hold water.
- To identify favorable sites for the construction of artificial recharge sites for ground water.
- To integrate information derived through remote sensing and non-remote sensing methods through GIS techniques.

In particular, any watershed management plans must be aimed at

- Purpose of the watershed management plan and intended use and objectives of the plan.
- Briefly describe the landscape to include a broad range of watershed characteristics and attributes impacting the planning process.
- Describe watershed in terms of physical and geographical landforms, hydrological, climatic and ecological features, land use land cover patterns, cropping history, total acres of crops etc.

- In the given watershed identify water-body use, classification and standards, fish and wild life resources, primary income source, i.e economic base, population demographics (density and distribution).
- Inventory of the resources in the watershed
- Describe the accepted or stated condition of the natural resources with in the watershed like surface water, groundwater, fish and wildlife, habitat / special ecosystems (wetlands), cultural resources and other resources.
- Identify and describe specific water resource management problems by analyzing the data and identifying the sources and causes of impairment. For example, point and non-point sources, physical and chemical pollutants, sediment load, degraded physical habitat, hydrological modifications, endangered species, agricultural pesticides, local concerns and standards.
- Formulate alternate management strategies by considering and evaluating alternatives to address the environmental problems.
- Use of appropriate best management practices to protect the watershed, identify any proposed land use changes, and identify permanent or structural solutions considered.
- Preparation of **Watershed management draft** by stating project goals and objectives
- Describe planned improvements in the watershed; establish the project milestones as identified by the local project participants. Review the draft plan with local stakeholders and make any modifications to gain acceptance from the community by convincing them that implementation of planned practices will adequately address the water resource problems.

The study includes :

Remote sensing data acquisition and analysis IRS-ID AND LISS data,

Hypsometry [Area altitude analysis]

Estimation of sediment yield

Erosion intensity zoning

Socio-economic data collection and analysis

Integrated study using GIS, by creating spatial as well as attribute databases

Action plan for land and water resources.

The database can be created through :

Digitization /scanning

Storing, editing, cleaning and labeling

Grid cell creation

Application of overlay

Methodology

The methodology of the study comprises of generation of thematic maps on 1:50,000 scale showing current land use / land cover, types of waste lands, forest cover, ground water potential zones, geology [rock types, structural features], soil types, hydrogeomorphology, drainage, slope etc.

Satellite images [FCC], IRS-IA LISS II A and Landsat TM, are used to extract, thematic topographic data. Thematic maps are prepared from remote sensing data validated with ground survey. These derived data systems are converted into consistent map formats using GIS software like Arc Info. Specific integration and analysis of these data can be performed to derive useful outputs in the form of maps and statistical data. This multi-thematic information is integrated manually using GIS techniques. The GIS techniques help to visualize. Organize, combine, analyze, predict and query the spatial data along with non-spatial data.

Integrated area planning in the study area is essential for maximizing benefits by means of proper management of natural resources available for sustainable development. In order to achieve this objective, the required study is divided into three phases.

PHASE – 1

Natural resources mapping and assessment.

It is done in 3 steps :

- Preliminary interpretation of satellite geocoded data
- Verification of doubtful areas that may arise during the first step, by conducting fieldwork [ground truth analysis]
- Finalization of thematic maps and preparation of maps using cartographic aids [digitization]

PHASE – II

Collateral data collection and analysis

PHASE – III

Integration of various data types through GIS techniques.

Data Source

- IRS SATELLITE ID PAN and LISS –III
- PAN-0.5-0.75
- LISS-III bands B2 [0.52-0.59μm]
    - o B3 [0.62-0.68μm]
    - o B4 [0.77-0.86μm]
    - o B5 [1.55-1.70μm]
- LISS-III GEOCODED
- PAN GEOCODED

- PAN+LISS-III MERGED
- MAPPING THROUGH –VISUAL INTERPRETATION-CONVERGING EVIDENCE

PHASE - I

- Collateral data collection and analysis
- Slope
- Transport network
- Drainage

(Slope, Transport network, Drainage: Using survey of India toposheet)

- Temperature
- Humidity
- Rainfall
- Socio economic factors
- Hypsometry- [area –altitude analysis]
- Drainage morphology
- Determination of intensity of the watershed
- Zones of erosion
- Land use Land cover patterns

PHASE-III

Integration of data using GIS

GIS database creation

- Spatial database creation
- Non spatial or attribute data creation
- Linking spatial and non-spatial database
- Topology creation

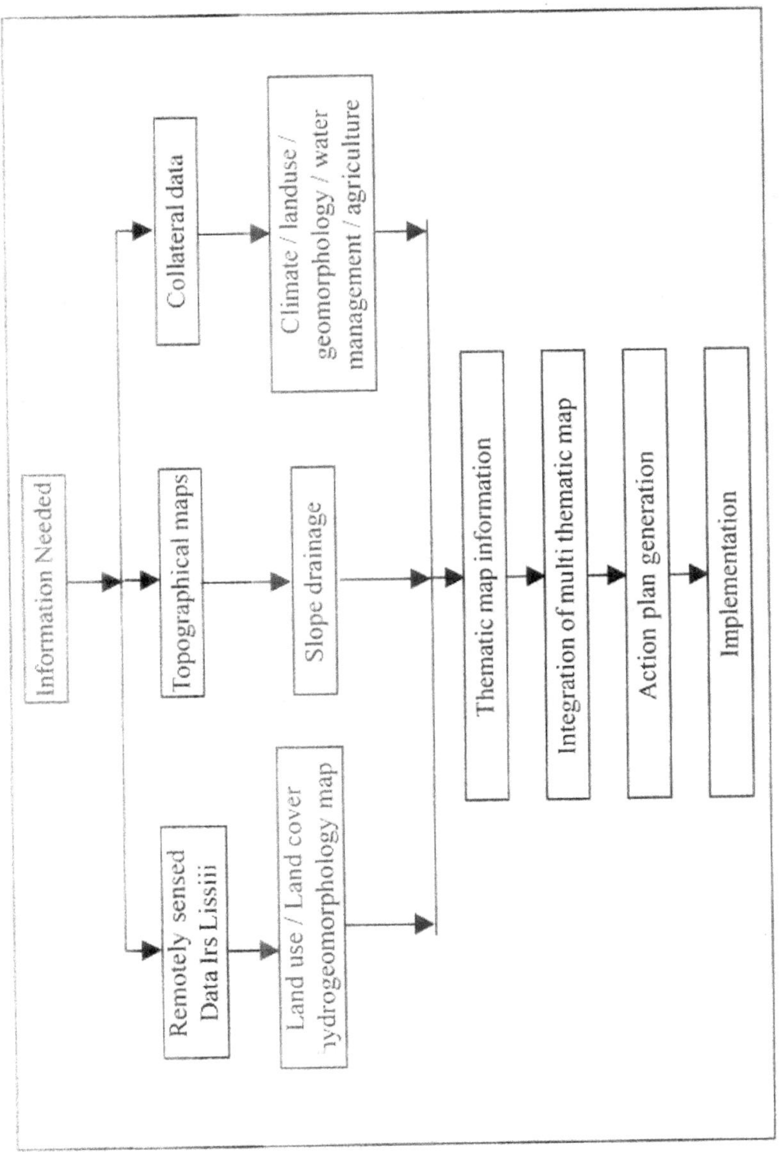

Fig. 12.1 Scheme of integrated watershed management

Spatial Database

- It represents geomorphological location of features.
- It can be represented on the map as a point feature or a line feature or/ and area features.
- Point features are used to represent the location of geographic phenomena at a point or to represent a very small feature.
- Line features are used to represent those features that are too narrow to be shown as an area or having no width.
- Area features are used to represent a region enclosed by line features.

Master action plan can be developed for water resources development in combination with socio economic factors of the study area.

Developments activities comprise of :

- Soil and water conservation measure in most of the area commencing from water boundary upto command area
- Forestry and posture development along the ridges and steep slope
- Dry forming practices in the rain fed areas of the watershed.
- Identification 9of the sites for the construction of the minor irrigation tanks across minor streams and valleys
- Construction of artificial recharge structures (percolation tanks , check damns where ground water is available.
- Identification of areas for dairy development where fodder is available.

Advantages of Integrated Studies

- Watershed management through integrated study by remote sensing and GIS can lead to better appreciation from the government and public of the benefits to formers. Maximizing benefits and achieving them quickly will convince the decision-makers of the validity of the approach. give a comfortable position in evaluation.
- Once water shed is properly managed, the local people in the rural area will live in better conditions. Hence, for the betterment of living conditions of rural people for their sustainable development, watershed development can be planned in an integrated manner by taking one watershed after the other.

13

Paradigms of Water Resources Development

U. Aswathanarayana

Mahadevan International Centre for Water Resources Management, Hyderabad, India.

(e-mail: uaswathanarayana@yahoo.com)

"Highest good is like water … because water excels in benefiting myriad creatures without contending with them, and settles where none would like to be, it comes close to the way …. In the world there is nothing more submissive and weak than water … Yet for attacking that which is hard and strong, nothing can surpass it … Tasteless, it accepts all tastes, colourless, all colours, reflecting the sky, refracting the white stones of its bed, dissolving or suspending the soils and minerals over which it flows. The pulse of the bodies is liquid, as indeed all living pulses are …." ——

Sayings of the Chinese philosopher, Lao Tzu, as quoted and annotated by Vikram Seth, in his book, **From Heaven Lake** (1983, p. 165).

Key words : Economic dimensions, Water withdrawals, Water transfers, Groundwater protection, Bio-pond technologies, Geopurification, Irrigation management, Planning of water supplies, Computer-aided water management, Recommendations.

Introduction

This chapter is largely drawn from the volume entitled, "*Water Resources Management and the Environment*" (Aswathanarayana, 2001).

All life is critically dependent upon water the origin and evolution of life became possible on earth because of the presence of liquid water. Animals and plants consist mostly of water. Water (40 L) constitutes three-fourths of the human organism.

In his keynote address to the UNESCO-organized World Conference on Science, Budapest, Hungary, June 26 - July 1, 1999, Dr. M.S. Swaminathan of India gave his vision of a hunger-free world. He envisages three approaches : (i) improving food production through the optimization of various inputs, via people-participatory technologies in an ecologically-sustainable manner, (ii) increasing access to food by promoting job-led economic growth, and (iii) facilitating the availability of safe drinking water and environmental hygiene, in order to ensure the biological absorption of food in the human body. It may be noted that water resources management happens to be at the core of all the three approaches. Plants need tremendous amounts of water to grow about 30-40 cm of water is needed to produce a food crop, such as wheat. Thus irrigation will continue to be the principal user of water.

Economic Dimensions of Water Use

The International Conference on Water and Environment, Dublin, 1992, enunciated two crucially important guiding principles, namely, that all human beings have a basic right to have access to clean water and sanitation at an *affordable price*, and that water has an economic value in all its competing uses and should be recognized as economic good.

The economic dimensions of the water use are summarized as follows in terms of location of water use (Table 13.1)

Table 13.1 Economic dimensions of the water use in terms of location of water use

	On site	In stream	Out stream (withdrawn)
Agriculture, Forestry, Aquaculture	Rainfed agriculture, Forestry, Swamps and wetlands habitat Utilization of estuaries	Fish and wild life Utilization of Estuaries Waste disposal**	Livestock Irrigation*
Industry		Hydropower Waste disposal**	Stream power Mining ** Cooling Processing ** Hydraulic transport
Infra-Structure		Navigation, Recreation Aesthetic enjoyment Waste disposal **	Drinking Domestic uses Public uses in Settlements

* Highly consumptive use; **Heavy impact on water quality

Water is a mobile resource which an be used and reused many times (e.g. municipal water supply → waste water → bio-ponds → irrigation water → groundwater recharge → withdrawal from wells, etc.). Since many combinations of use and reuse of water are possible simultaneously or in sequence, it is critically important to know how the various uses of water combine and interact in space and time. Consideration has therefore to be given to return flow and reuse. A given physical unit of water can be used over and over again if water is used fast and returns large quantities of clean or well-treated water. Even though the value addition for each use of water is small, a pattern of multiple use can generate large values. The location and timing of water use becomes important in the system context of water use. The system gains more value when the waste-generating users are located as far downstream as possible.

Water use has to be optimized, using pricing as a control. Also, research and development is leading to more efficient use of water, use of inferior quality of water and reuse and recycling of water. Thus, instead of investing more money in developing new supplies of water, it is generally cost-effective to use the same money to develop methods of efficient use of water and recycling and reuse of water.

Allocation of Water Among Competing Users

A case history is given to illustrate the use of linear programming methodology to decide about the allocation of water to competing users.

Barandemaje tried to develop a policy aimed at optimizing the allocation of water to competing industrial users in Burundi (Africa). A finite amount of water (26,000 m^3 /d) has to be allocated to five competing industrial users (a cigarette company, a brewery, a textile mill, a paper mill and a dairy). A linear mathematical model was developed to conceptualize the five firms. It consists of an objective function and constraint equations. Each equation takes into account the water-use / net-profit relationship for each firm. The constraints are based upon the lower and upper water consumption capacities and the non-negativity conditions on the value of the decision variables. Optimization methods were applied to two kinds of situations: involving only competitive criteria (i.e. which kind of allocation would yield the maximum profits), and involving both social criteria (the needs of the dairy being given the highest priority and the cigarette company, the lowest) and economic competitive criteria. Even though the dairy may provide less profit per unit of water used relative to the cigarette company, the community needs milk more than it needs the cigarettes, and hence the higher priority for the dairy in water allocation. On the basis of this kind of analysis , the decision-makers could make their choices.

Water Withdrawals

The sectoral water withdrawals, by country income group are summarized in Table 13.2 (source: World Resources Institute, 1990):

Table 13.2 Withdrawals by sector for different income groups

Income group	Annual withdra-wals per capita (in m³)	Withdrawals by sector (% and m3)		
		Domestic	Industry	Agriculture
Low-income	386	4 % (15 m³)	5 % (19 m³)	91 % (351 m³)
Middle-income	453	13 % (59 m³)	18 % (82 m³)	69 % (313 m³)
High-income	1167	14 % (163 m³)	47 % (548 m³)	39 % (455 m³)

($1\ m^3\ y^{-1} = 2.74\ L\ d^{-1}$)

The data in Table 13.2 allows us to draw the following conclusions : (i) the total withdrawal of water by the low-income groups is less than one-third that of the high-income groups, (ii) the low-income groups use most of their water (91 %) in agriculture (evidently inefficiently, as they are unable to grow adequate amount of food for themselves), (iii) the use of water in the industry by the high-income groups is about 30 times more than that by the low-income groups, and (iv) the quantity of water used for domestic purposes by the low-income groups (about 40 L d^{-1}) is less than one-tenth (about 450 L d^{-1}) of that used by the high-income groups.

The very limited quantity of water that the low-income groups use for domestic purposes, cannot but have profound health consequences. More than one billion people in the Developing countries are still without access to safe water and about 1.7 billion people are without access to adequate sanitation facilities. This has disastrous health consequences. About 900 million cases of diarrheal diseases, resulting in the death more than 3 million children, are attributed to the lack of safe water and proper sanitation. At any time, 200 million people are suffering from schistosomiasis or bilhazaria, and 900 million from hook worm. Other water-related diseases like cholera, and typhoid are rampant. During 1997-98, the cholera epidemic caused by El Nino -induced flooding in East Africa killed an estimated 5000 people. Apart from human suffering involved, these illnesses cause enormous economic losses.

Water Transfers

Water as an input may be transformed into a variety of outputs. As water becomes more and more scarce, the value of the possible outputs will have to be taken into account in allocating the water to the competing uses. In USA, it has been estimated that one acre-foot of water typically yields only about USD 400 on the farm versus USD 400,000 in manufacturing. In such a situation, it is inevitable there would be transfer of water from lower economic returns (e.g. agricultural) to higher economic returns (municipal and industrial).

An innovative system of water transfer is water banks that have been set up in California. Just as we put money in the bank, and the bank lends it to a borrower on interest, farmers who have water to spare will place their water at the disposal of the bank. The bank will then sell the water to the needed persons / communities / industries, and pay the farmer. During the California drought of 1987-1993, the water bank was handling water at the rate of about a billion m^3 per year.

Drinking Water

The daily physiological consumption of drinking water for humans varies from 1-4 L per capita per day, depending upon the climate (high in the arid regions), the kind of work a person does (a manual worker working in the open sun would need to drink more water, than a person working in an air-conditioned office), and social habits. The Tuaregs who live in the 130^0 F heat of the Sahara, sweat about 8 L of water a day unless the liquids lost are replenished, a person would suffer dehydration, leading to coma and death (source : *National Geographic*, Mar. 99).

Animals need larger amounts of drinking water : 40 L/d for cattle, horses and camels, 15 L /d for hogs, and 10 L/d for sheep. As should be expected, the largest land animal, the elephant, consumes huge quantities of food (about 300 kg of plant matter per day) and water (about 90 L/d, the same as 6 giraffe and 66 impala). Animals can consume semi-potable water without any harm. For instance, the limit for Total Dissolved Solids (TDS) for potable water for human consumption is 500 mg/L. Water with TDS of, say, 700 mg/L can be safely given to a camel. In the process of evolution, animals in the wild have acquired the ability to survive on small quantity of very poor quality water. For instance, an impala can survive on 1.5 L/d, whereas a sheep needs 10 L/d. So in areas of South Africa, where water is scarce, wild animals, such as impala and ostrich, are reared in farms for meat.

Monitoring of Drinking Water Quality

The parameters of water that are to be routinely measured are the following:

Physical parameters **:** Turbidity, Suspended solids, Electrical conductivity, Colour, Odour.

Chemical parameters :

(i) pH, carbonate, chloride, sulphate, nitrate, Total Dissolved Solids, ammonia, alkalinity, hardness, BOD, COD, etc.

(ii) Fifteen trace elements (as recommended by IMEP-9 of the European Commission) : B, Ca, Cd, Cr, Cu, Fe, K, Li, Mg, Ni, Pb, Rb, Sr, U, Zn, and F.

Biological parameters :

Total coliforms/100 mL. Also as many pathogens as possible from among the following: *Campylobacter spp., Giardia Lamblia, Entamoeba histolytica, Shigella spp., Vibrio cholera, Salmonella typhi, Salmonella spp., Escherichia coli (path.), Enteroviruses, Hepatitis A virus, Ancyclostoma duodenale, Trichuris trichuiria, Taenia saginata, Ascaria Lumbricoides, etc.*

Groundwater

Quality of Groundwater

The importance of groundwater arises from the following considerations: (i) The quantity of groundwater upto drillable depth, is about seventy times more than all the waters in the rivers, lakes, reservoirs, etc. in the world put together, (ii) the quality of groundwater is generally superior to surface water, because the soil column purifies the contaminants in water through processes such as anaerobic decomposition, filtration, ion exchange, etc. (iii) Groundwater is the main source of potable water in most parts of the world (for instance, groundwater is the source of 75 % of the municipal water supplies in USA). The use of groundwater for irrigation ranges from about 40 % in the case of India, to about 70 % in the case of Libya. As a consequence of excessive use of groundwater for drinking and irrigation purposes, the water table is going down 1-3 m per year in most parts of India.

The spatial and temporal distribution of groundwater quality has been found to be dependent upon climate (precipitation and evaporation), topography (slope gradient, which determines the residence time of groundwater), geology of the area (mineralogical and chemical composition of the rocks and soils with which groundwater is in contact), etc.

If the precipitation exceeds potential evapotranspiration for ten months in a year, the groundwater is recharged virtually all the year round. The water table generally follows the topography. As the topography is undulating, the groundwater moves fast. As the reaction time of the groundwater with rock is short, very little leaching is effected. Consequently, the groundwater has low electrical conductivity (100-450 lS cm^{-1}), and low concentration of major ions. Surface water stored in tanks is characterized by very low electrical conductivity and low concentration of ions. Since groundwater is recharged by seepage from tanks, its major ion concentrations are similarly low. As should be expected, the groundwater is of the calcium bicarbonate type.

Pollution of Groundwater

The types of activities that can cause groundwater contamination at various levels are summarized as follows (source: US EPA's Citizen's Guide to Groundwater Protection, 1990)

Surface of the ground : Infiltration of polluted surface water, Land disposal of wastes, Stockpiles, Dumps, Sewage sludge disposal, Use of salt in de-icing of roads (in cold countries), Animal feedlots (and manure heaps), Fertilizers and pesticides. Accidental spills. Airborne source particulates. Human and animal faeces, etc.

In the ground, above the water-table : Septic tanks, cesspools and privies, Holding ponds and lagoons, Sanitary landfills, Waste disposal in excavations, Leaks from underground storage tanks, Leaks from underground pipelines, Artificial recharge, Sumps and dry wells, Graveyards, etc.

In the ground, below the water-table : Waste disposal in wells, Drainage wells and canals, Underground storage, Mines, Exploratory wells, Abandoned wells, water-supply wells, groundwater withdrawal, etc.

The chemistry of groundwater is affected by a large number of factors, such as, physiographical, geological, physico-chemical, physical, biological, anthropogenic, etc.

The movement of the groundwater is generally slower and less turbulent than that of the surface water. Though chemical contamination of groundwater may sometimes be caused by natural sources (e.g. calcium-rich waters in a limestone country), most of the contamination occurs due to substances produced and introduced by man. These include synthetic chemicals (such as, solvents, pesticides and hydrocarbons), landfill leachates (i.e. liquids that have leached landfill dumps, and carry dissolved substances from them), organic wastes (e.g. bacteria and viruses from night soil dumps, hospital garbage), etc.

When contaminated water moves through the soil column in the unsaturated zone, the contaminants present in the water tend to be removed through processes such as anaerobic deposition, filtration, ion exchange, adsorption, etc. On the other hand, the content of dissolved solids may increase as a consequence of the reaction between soil and water. The velocity of movement of the contaminant through and into the groundwater depends upon the physical and chemical characteristics of the soil, the nature of the contaminants and the flow system of the zone. It may be highly variable - ranging from nearly instantaneous to hundreds of years. If the aquifer concerned is composed of coarse-grained material like gravels, water may travel quickly, with residence times of a few days or weeks, before it is abstracted or reaches a river. On the other hand, if the aquifer is composed of fine-grained material such as chalk, water will move slowly. Consequently, water may reside in such aquifers for years, or even centuries, before it is abstracted or discharged.

The manner of spreading of a pollutant in an aquifer depends upon the nature of the pollutant and the aquifer. If the pollutant concerned is soluble (e.g. chloride or sulphate salts), it will get dissolved in the groundwater, and will move as a plume along the same path and with the same velocity as groundwater. On the other hand, if the pollutant is insoluble or poorly soluble (e.g. chlorinated solvents), its behaviour will be markedly different from that of the soluble pollutant. The insoluble pollutant will tend to remain as a separate phase, and "may sink below the water table, and flow separately along low permeability layers encountered at depth in the aquifer, or be trapped by capillary forces and act as a long-term contaminant source".

Fig. 13.1 gives the groundwater model showing the disposition of the contaminant source and the resulting plume (source : van der Heijde, 1992).

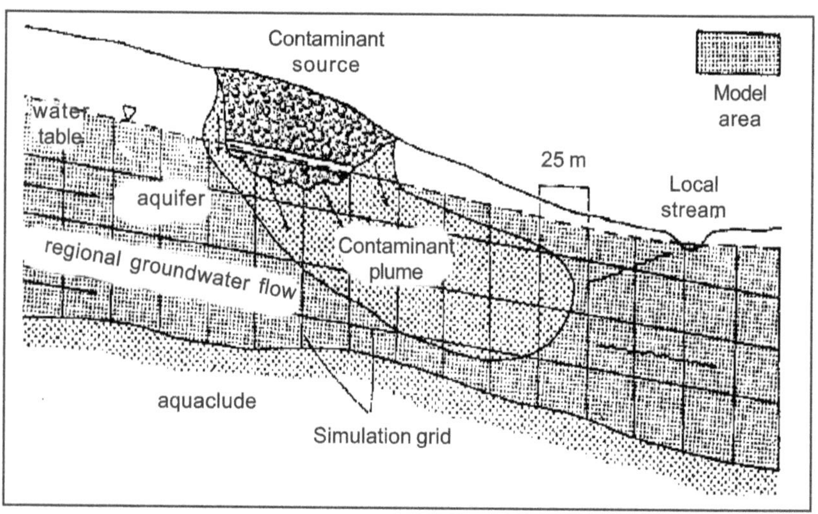

Fig. 13.1 Vertical cross-section of the groundwater model showing the disposition of the contaminant source and the resultant pollutant plume (source : van der Heijde, 1992)

The clean up of the an aquifer is technically complex, extremely costly and takes a long time to accomplish. But there might be circumstances where it may have to be undertaken nevertheless because of the present-day or future hazards (such as, pollution of drinking water wells, and seepage of the polluted groundwater into fresh surface water). An aquifer may have to be flushed many times (typically 5-10 times) in order to remove both solute and adsorbed pollutant.

Protection of Groundwater

The French hydrogeologist, Margat, pioneered the concept of groundwater vulnerability, and ways and means of addressing it through vulnerability maps (Albinet and Margat, 1970).

Groundwater originates principally through the infiltration of the precipitation falling on the land surface. Thus, many activities (domestic, agricultural, industrial, etc.) taking place on land have the potential to adversely affect the quality and availability of groundwater. An important consideration to be kept in mind is the slow response of most of the groundwater systems. Once pollution has occurred, the water has to be treated at the point of abstraction. As has been stated earlier, the cleanup of an aquifer is technically complex, extremely expensive, and takes a long time to accomplish. It follows that every effort should be made to prevent the contamination of the groundwater in the first instance. An ideal situation would be to manage the land solely to protect the gathering of groundwater (a few such areas continue to exist in Germany and England since historical times). In practice, this is just not practicable. We have to accept trade-offs between economic development and aquifer protection as a fact of life only, we should do this objectively. Groundwater protection strategies have to be designed by meshing hydrogeological understanding with the requirements of land-use policies, so as to lead to land-surface zoning (Foster and Skinner, 1995).

The protection policy is based on the following elements of aquifer pollution vulnerability: (i) We start with defining what activities could be permitted on land without entailing no more than acceptable risk to the groundwater, (ii) We take advantage of the fact that soils overlying a saturated aquifer have the capacity to attenuate the contaminants passing through the soil, and (iii) It may not be always possible or even necessary to give equal protection to the total available groundwater resource, such as an aquifer. Protection may be selectively given for potable and sensitive uses or specific sources, such as boreholes, wells and springs. Thus, Source Protection Areas (SPAs) may be earmarked around existing or projected sites of major potable water supply. (Fig. 13.2, source : Foster and Skinner, 1995).

Source protection (or wellhead protection as it is called in USA) is an essential part of groundwater protection. The objective of demarcating source protection areas or zones, is to protect the groundwater from two kinds of pollutants : (i) contaminants which decay with time - the longer they reside in the subsurface, the greater will be their attenuation, and the less will be their ability to contaminate the groundwater, (ii) non-degradable contaminants, whose adverse impact on the groundwater can be mitigated through flowpath-dependent dilution. Most of the hydrogeological environments can be modeled using the computer program, FLOWPATH. Where the hydrological complexity is greater, recourse may be taken to the use of the programs, MODFLOW-MODPATH -MODELCAD.

Water Quality in Relation to Water Use

Water quality gets modified in the course of movement of water through the hydrological cycle, through the operation of the following processes: evaporation, transpiration, selective uptake by vegetation, oxidation / reduction, cation exchange, dissolution of minerals, precipitation of secondary minerals, mixing of waters, leaching of fertilizers and manure, pollution, and lake / sea biological processes.

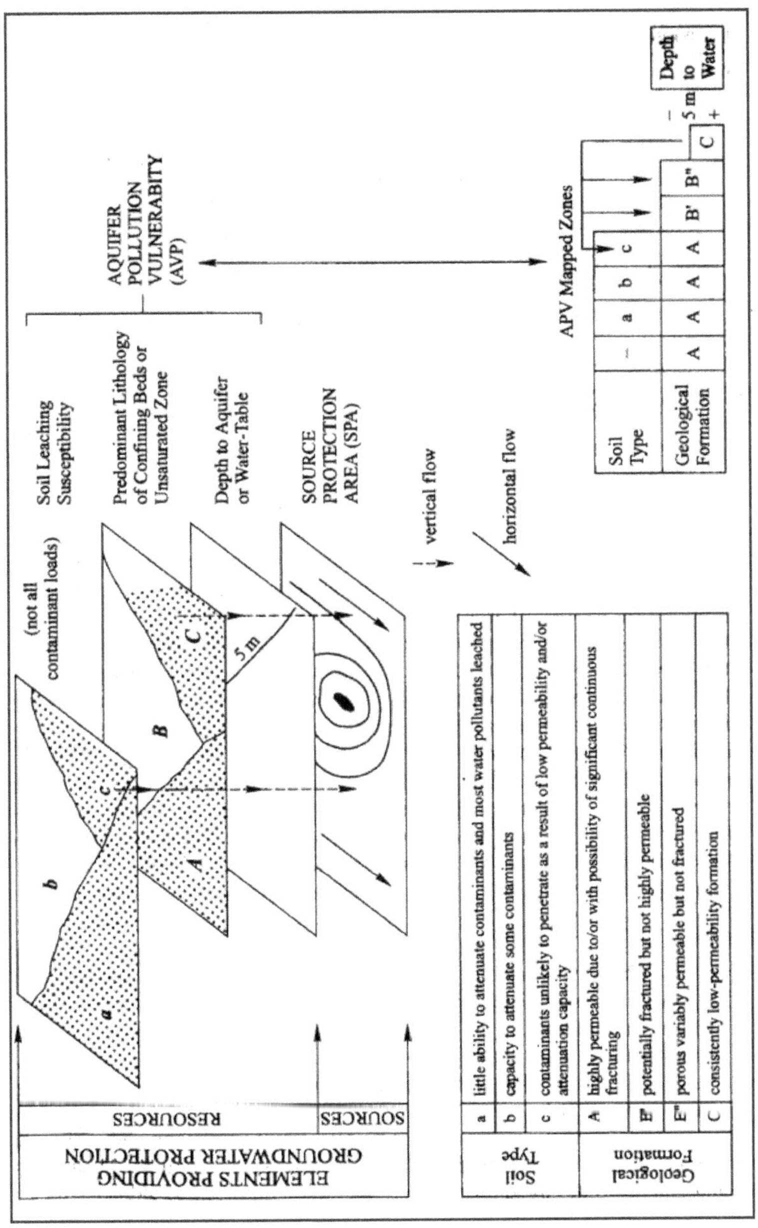

Fig. 13.2 Elements of groundwater protection and classification of aquifer vulnerability (source : Foster and Skinner, 1995).

The various uses of water can be classified into three categories, depending upon the stringency of specifications for a particular use :

Most demanding uses : A - Drinking water ; B - Fisheries ;

Less demanding uses : C - Recreation, D - Industrial uses, E - Irrigation,

Least demanding uses : F- Power and cooling ; G - Transport

How the water quality contaminants constrain the use of water is indicated in Table 13.3.

Table 13.3 Water quality constraints in water use

Pollutant	A	B	C	D	E	F	G
Pathogens	xx	0	xx	0 to xx (1)	x	na	na
Suspended solids	xx	xx	xx	x	x	x (2)	xx (3)
Organic matter	xx	x	xx	x to xx (4)	+	x (5)	na
Nitrate	xx	x	na	0 to xx (1)	+	na	na
Salinity (9)	xx	xx	na	x to xx(10)	xx	na	na
Inorganic toxics (11)	xx	xx	x	x	x	na	na
Organic toxics	x	xx	x	x to x (1)	x	na	na
Protons(acidification)	x	xx	x	x	na	x	na
Aquatic biota	x (5,6)	x (7)	xx (12)	x to xx (4)	+	x (5,13)	x (8)

xx : Presence of the pollutant markedly impairs the use of water for the concerned purpose ; can only be used for the purpose after major treatment ;

x : Minor impairment of use, needing minor treatment ;

0 : No impairment of use is involved ;

na : Not applicable;

\+ : Degradation of water quality in this way may be beneficial for this specific use.

1 : Food industries ; 2 : Abrasion, 3 : Sediment settling in channels, 4 : Some industries (e.g. electronic industries) require low DOC, 5 : phytoplankton may cause clogging of filters, 6 : Bacterial activity may affect odor and taste, 7 : Higher algal biomass is acceptable in fish ponds, 8 : Development of water hyacinth (*Eichhornia crasspipes*), 9 : Also includes B and F, 10 : Ca, Fe, Mn in textile industries, 11: As, Cd, Cu, Hg, Pb, Se, Sb, Sn. Zn, etc. 12. Loss of transparency from algal development, 13. Development of zebra mussels in pipes (*Dreissena polymorpha*).

The information provided in Table 13.3 (source: Meybeck, 1998) indicates the directions for proper use and reuse: (i) The highest quality water (at the rate of, say, 2-4 L per capita per day) should be reserved for drinking purposes (as different from water for other domestic uses, such as bathing, washing clothes and utensils and sanitation, which can use slightly inferior water). In many Indian homes, the traditional practice has been to keep drinking water separately from water to be used for washing purposes. It is no doubt technologically feasible but prohibitively expensive to purify contaminated water (say, municipal waste water) to a level of purity acceptable for domestic purposes (ii) Municipal waste water which contains (say) pathogens can be used in most industries, except food industries, (iii) Sewage which contains organic matter and nitrates is not only acceptable but even be desirable for use in irrigation.

Augmentation and Conservation of Water Resources

Rainwater Harvesting

Any method that saves water is a conservation method. Since ancient times, farmers and herders living in areas of limited rainfall, practiced a variety of water conservation measures. This allowed population densities of 10-500 persons/km^2 in areas with annual precipitation of 100-150 mm. The traditional techniques of water conservation can be described under three heads: (i) making better use of rainfall by minimizing runoff losses (in combination with increased filtration), collection and storage of rainfall, minimizing evapotranspirational losses, and improving rainwater usage by plants, (ii) collection of fog drip and dew, and (iii) making use of groundwater without water lifting, through quanats, artesian wells and horizontal wells. These water conservation measures are low-cost, labour-intensive operations and are hence particularly suitable for the Developing countries. It should, however, be borne in mind that rainwater can be harvested only if there is rain no rain, no rainwater harvesting !

Rainwater harvesting can be done on any kind of roof. Roofs of corrugated galvanized iron sheets are common in most Developing countries they are cheap and durable and have low maintenance costs. They tend to rust in the coastal regions, however. The poorer people live in mud huts with thatched roofs. In the case of thatched roofs, the rainwater would tend get colored, and therefore unattractive. Also thatched roofs are less durable than corrugated iron sheets or tiles. One way to get over the problem is to cover the thatched roof by PVC sheet during the rainy season of about four months. Alternatively, bituminous paper or sisal or coir-cement corrugated sheets could be used. Metal or bamboo guttering may be used. The first flush of rainwater is not collected, as it may be carrying dust, worms or unwanted debris. The harvested rainwater is made to pass through a baffle or filter to remove any impurities. The harvested water may be stored in a variety of containers, depending upon what a family can afford "ghala" tank, brick structure, or ferrocement tank, aluminum tank, etc. A ghala tank is a large granary basket. It is made by molding cement inside and outside of a bamboo framework. If the cement wall thickness is 3 cm., the ghala tank could have a capacity of 2500 L. The usual practice is to keep a tap to draw water about 15 cms above the bottom of the container to avoid mud that may accumulate at the bottom .

Harvested rainwater is protected from contamination by a simple device, namely, a baffle tank. The runoff passes through a baffle and sieve, before going over a stand-pipe. Sediment settles down in the bottom of the box, which needs to be periodically cleaned .

A family can be assured of clean water for domestic use at the rate of about 60 L/d, assuming a rooftop with an area of 50 m^2, rainfall of 1000 mm, and collection efficiency of 50 %. Bermuda is an excellent example of efficient application of rainwater harvesting technology. Bermuda depends for its supplies of water (80 L per day per capita) almost exclusively on harvesting of precipitation (1430 mm.). The rainwater is collected and stored in cisterns beneath the houses. Necessity, tradition, financial incentives and administrative procedures have been effectively integrated to bring about this highly desirable situation. The Portuguese colonizers promoted rainwater harvesting in some towns in Mozambique (e.g. Pemba) by incorporating the facility in the building design. Rainwater harvesting is practiced in Israel and in the Rajasthan province in India.

There are some simple methods of conserving water in homes. Bathrooms can be so designed that the wastewater from bathing and washing passes through a filter to a cistern from where it can be used to flush the lavatories by gravity flow. Indian-style, pour-flush lavatories with stainless-steel commodes require just 2.5 L of water per flush, as against 15 L required for normal W.C.s. Besides, wastewater from bathing and washing is often used to water the gardens, which tend to have a preponderance of low water need plants (such as, bougainvillea, cactus, roses, etc.).

Farming communities in arid lands (e.g. South Africa) routinely save wash water and give it to animals, such as pigs. Micro-enterprises are ideally suited to promote rainwater harvesting by individual families in the Developing countries.

Harvesting of Surface Runoff

As in the case of rainwater harvesting, collection, storage and distribution of storm runoff, have been practiced by ancient cultures from time immemorial. In South India, thousands of irrigation tanks were built by the ancient kingdoms, several of which continue to be in use to this day (currently, minor irrigation tanks in India have a storage capacity of about 14.3 M ha m).

In Negev desert, rainwater from rooftops, courtyards and streets is collected into settling tanks. Prior to storage, water collected from ground catchments is cleansed by the use of silt traps and sand filters. Clear water is then channeled into underground cisterns in the house. The capacity of the cistern, which may be in the form of well-sealed, brick and mortar construction, may be 5-10 m^3. The annual water requirements of a family are usually computed at the rate of 1.5 m^3 per person, and 1 m^3 per camel. Aboveground storage tanks with a capacity of 1000 L could be built with galvanized iron sheets, but when full, such a tank may get deformed, unless reinforced by steel or wooden framework.

Diverting the flood for warping disperses the runoff and moderates the flood peak. Thus warping reduces the flood hazard, besides converting a barren gully or flood plain into a fertile cropland. The usual practice is to construct a dam at the mouth of the gully to divert the muddy water to the flats on either bank of the stream. In some localities in India, flood water is diverted for leaching saline soil.

The usual practice is to construct a dam at the mouth of the gully to divert the muddy water to the flats on either bank of the stream. Alternately, intercepting ditches and canals can be constructed along contour lines to intercept the overland flow on hill slopes, and the water thus intercepted is used for warping river flats and terraced fields (Fig. 13.3; source: UNEP, 1983).

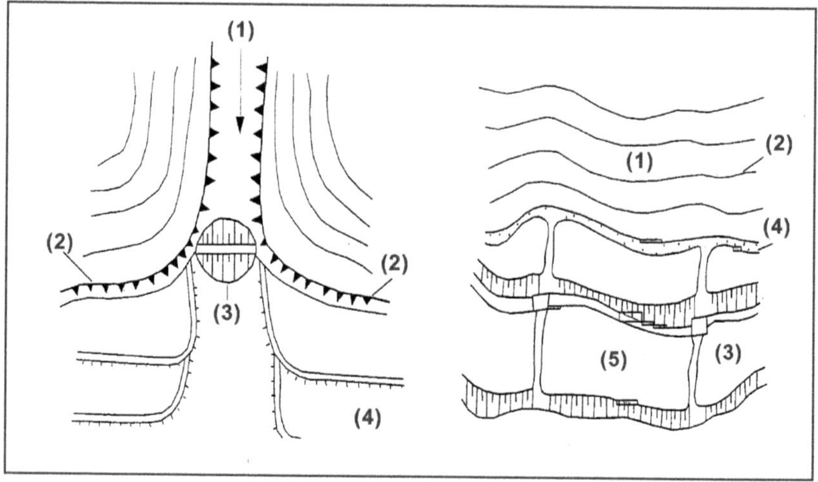

Fig. 13.3 Construction of the ditches and canals along the contour lines for intercepting the surface runoff

(Source : UNEP, 1983)0

Storage and utilization of rainwater from ground catchments

Rainwater storage ponds are best located where there is (i) a natural, saucer-shaped depression, or (ii) a shallow ravine with a small flowing stream which can be impounded, and (iii) the ground has high water retention capacity. In Australia, self-contained butyl bag installations with a capacity of 228,000 L are commercially available for providing water to the livestock.

If the soil porosity is high, the bottom and the sides of tank have to be sealed with a suitable sealant such as an embedded membrane of plastic sheeting or bentonite. Powdered sodium bentonite is spread over the dry surface at a minimum rate of about 5 kg/ m^2. Soils with high sand content need to be compacted, before bentonite is applied at the rate of 19.6 kg/ m^2. In Hawaii, the usual practice of sealing is to use top and bottom layers of butyl laminated with 0.8 mm nylon.

The various techniques of reducing evaporation by covering the surface of rainwater storage basins, are summarized below (Table 13.4 ; source : UNEP, 1983).

Table 13.4 Techniques of reducing evaporation

Technique	Reduction in evaporation (%)
1. *Changing of water color*	
Dye in the water	6 - 9
Shallow colored pans	35 - 50
2. *Using wind barriers*	
Baffles	11
3. *Shading the water surface*	
Plastic mesh	44
Blue, polylaminated plastic sheeting	90
4. *Floating reflective covers*	
White butyl sheets	77
Polystyrene rafts	95
Continuous wax	87
Foamed butyl rubber	90

A low-cost, ecologically-sound option would be to line the pond with clay (to reduce seepage), and cover the pond with bamboo or wooden poles and train viny vegetables with large leaves to cover the pond. The vines would not only reduce evaporation but would also provide vegetables.

Sometimes, groundwater may exist under artesian conditions, and may be able to flow out under its own pressure (i.e. without pumping). There may be situations where the groundwater is trapped by a vertical impervious geological barrier. Such groundwater can be tapped by boring a horizontal hole, and inserting a steel pipe casing. Horizontal wells are inexpensive to construct, and no pumping is needed. There is no possibility of contamination by rubbish or animals.

There are 500,000 tanks in peninsular India for providing drinking water and irrigation. In the province of Karnataka alone, tanks have a storage capacity of 200,000 m^3. But soil erosion and consequent silting is reducing the storage capacity of the reservoirs at the rate of 0.5 % per year. It is therefore critically important to minimize silting of reservoirs through practices such as the afforestation of the catchment, banning of cultivation of the foreshore lands, construction of soil traps in the catchment, etc.

Suspended sediment constitutes the chief visible impurity in the case water impounded by earth dams. Where water is especially precious, as in Eyre Peninsula in Australia, livestock is not allowed direct access to reservoirs, but is allowed to drink only from the troughs. A simple float-controlled outflow system from an earthen hillside dam taps water just below the surface but away from the muddy sediments.

Pot filters are used in homes to remove turbidity. In most cultures, indigenous plant products or natural products are used to remove turbidity from drinking water.

Groundwater Recharge

There are several advantages in storing water in underground reservoirs : (i) water in an underground reservoir is better protected against pollution than water in the surface reservoirs, (ii) underground reservoirs do not interfere with the land use, as the surface area above the underground reservoir can still be used for economic purposes, such as agriculture, (iii) water stored in the underground reservoirs is not lost due to evaporation (which may correspond to a layer of about a metre for surface water bodies in tropical countries), (iv) though the recovery rate of the water stored in an underground reservoir is less than unity because of effluent seepage of groundwater, that is beneficial as it will feed the rivers and makes for an increase in stable runoff.

A sensible solution is to go in for a combination of surface reservoirs and artificial injection, particularly where the recharging has to be done with flood water. The spring flood flow in the temperate regions usually lasts for one month, and rarely as long as three months. The monsoon flood flow in the tropical lands usually lasts 3 to 4 months. Thus the underground storage of the flood water needs to be accomplished in a few months' time. By having a combination of above-ground and below-ground reservoirs, groundwater recharge could be done throughout the year.

An excessive use of runoff water will reduce the groundwater runoff and consequently the stable part of the runoff. Groundwater moves slowly, and therefore has a high regulating capacity. Thus, a good practice would be to charge the groundwater during wet seasons and years, to be used when needed.

Wastewater Reuse Systems

Why Reuse Wastewater

Most developing countries tend to discharge the wastewater either untreated or partly treated, into the hydrological system. This has disastrous consequences. *Every cubic metre of waste water discharged into water bodies or water courses contaminates and degrades 8 to 10 cubic metres of good water.* Hence, prime waters get polluted and become non-potable. Thus most parts of the world are already facing the problem of degradation of quality of water because of the inherently wrong practice.

The waste water problem has indeed a win-win solution. Ecological treatment of waste water has a number of merits: besides removing the pollutants effectively, it is low-cost, energy-saving, resource-recovering, easy operation and maintenance. The reuse of wastewater not only avoids contamination of the hydrological system but the treated water can be put to beneficial uses such as irrigation, industry, recreation, etc.

Bio-pond treatment of wastewater

The Chinese experience in the biological methods of treatment of wastewater, is very relevant to all developing countries. In 1988, China had 86 conventional sewage treatment works with a total capacity of 3 M m^3/y, accounting for about 11 % of the municipal sewage flow in the whole country. As the country did not have the requisite financial and technical resources to build conventional sewage treatments for the rest 89 % of the municipal sewage flow, they went in a big way for eco-pond technologies.

Fig. 13.4 gives the scheme of the biopond treatment of the waste water (source: Wang,1991).

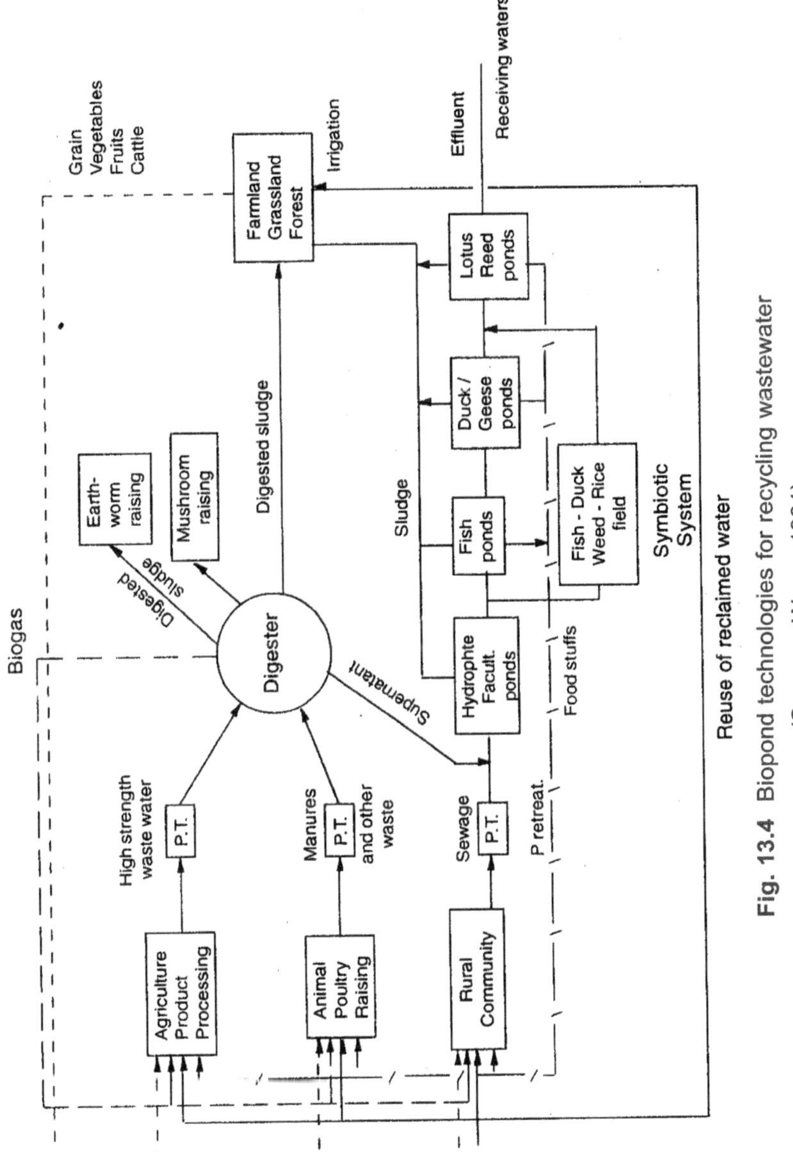

Fig. 13.4 Biopond technologies for recycling wastewater (Source : Wang, 1991)

Unlike the conventional wastewater treatment plants, which are energy-intensive (for pumping, stirring, filtration, reverse osmosis, etc.), eco-ponds are powered by solar energy, which is free. Eco-ponds serve the following useful purposes : (i) They are able to remove SS, BOD, COD, nutrients such as N and P, refractory organic compounds, bacteria, viruses and mineral salts, (ii) They use the food webs involving producers (such as algae and aquatic plants), and decomposers (such as, bacteria and fungi) to provide food to the consumers (such as, fish, clams, duck and geese), biogas fuel and organic fertilizer, and (iii) The effluent from the eco-ponds is sufficiently pure for it to charge the groundwater, or for non-potable uses (such as, irrigation, industrial, recreation, etc. uses). China plans to spend USD 36.2 billion during the next ten years to equip all the Chinese cities with wastewater treatment plants.

Using the pretreated wastewater for fish farming has become lucrative in all parts of China. The high rate of fish production (4,000 to 16,000 kg/ha) with very low cultivation cost makes fish farming more attractive financially than raising rice or wheat or vegetables. The practice of raising paddy and fish together has a long tradition in China dating back to Han Dynasty (206 B.C. to 220 A.D.). The rice plant benefits the fish by the intake of substances (e.g. excrement) that are harmful to the fish. As the fish (grass carp) swim through the water in search of food, this activity promotes aeration by breaking up the algae film and stirring up the upper soil layer. The grass carp excrete 72 % of its diet - so much so, 6800 grass carp can produce 1650 kg of manure per ha, which is more than enough to fertilize the paddy field. When young, the grass carp eat the abundant zooplankton that is present in the sewage sludge or supernatant from the digesters. As they mature, they eat weeds and the lower leaves of the rice plants. Also, the grass carp can kill more than 90 % of the mosquito hatchings in the paddy field.

The practice of raising rice and fish together is very much in vogue in eastern India and Bangladesh.

Geopurification

Geopurification or Soil - Aquifer - Treatment (SAT) is a low-technology treatment system, whereby the sewage effluent is subjected to groundwater recharge with infiltration basins. After SAT, the water will have very low suspended solids content, essentially zero BOD, much reduced concentrations of N, P, organic compounds and heavy metals, and almost zero levels of pathogens.

Four types of Soil-Aquifer Treatments are possible: (A) natural drainage of renovated water into stream, lake or low-lying area, (B) Collection of renovated water by subsurface drain, (C) Infiltration areas in two parallel rows and lines midway between, and (D) Infiltration areas in centre surrounded by circle of wells.

Settled sewage (primary effluent) may be used on industrial and fodder crops, pastures and hay, vegetables eaten in a cooked state, fruit trees, ornamental plants or seed plants. Irrigation should be stopped one month before the harvest in the case of apples, pears and plums, and furrow-irrigated broccoli and cauliflower, and tomatoes used for canning.

Use of Water in Agriculture

Water requirements of crops

There is much variation in the needs of water for crops. For instance, one hectare of sugarcane needs 300 ha.cm of water. The same amount of water could be able to irrigate about 8 ha of wheat or 30 ha of bajri (pearl millet) (Fig. 13.5; source: Falkenmark et al, 1990)

This statement has profound implications: (i) farmers need to be charged *pro rata* to the amount of water consumed, but not on the acreage of the farm, for the simple reason a farmer growing one ha of paddy uses lot more water than a farmer growing (say) one ha of sorghum, (ii) In areas of water scarcity or during times of drought, only low-water need crops, like maize and millet, should be allowed to be grown, even if a particular farmer has enough water resources to grow sugarcane or paddy.

Water has to be delivered to the plant in tune with the rooting depth of the crop, and available water capacity of the soil (150 - 200 mm per meter-depth of the soil for clayey soils, 100 - 150 mm for loamy soils, and 50 - 100 mm for sandy soils). The rate of water uptake by the plants depends upon (i) characteristics of the plant (rooting density, rooting depth, rate of root extension, etc.), (ii) properties of the soil (water retention and conductivity), and (iii) weather conditions (which determine the quantum of transpiration).

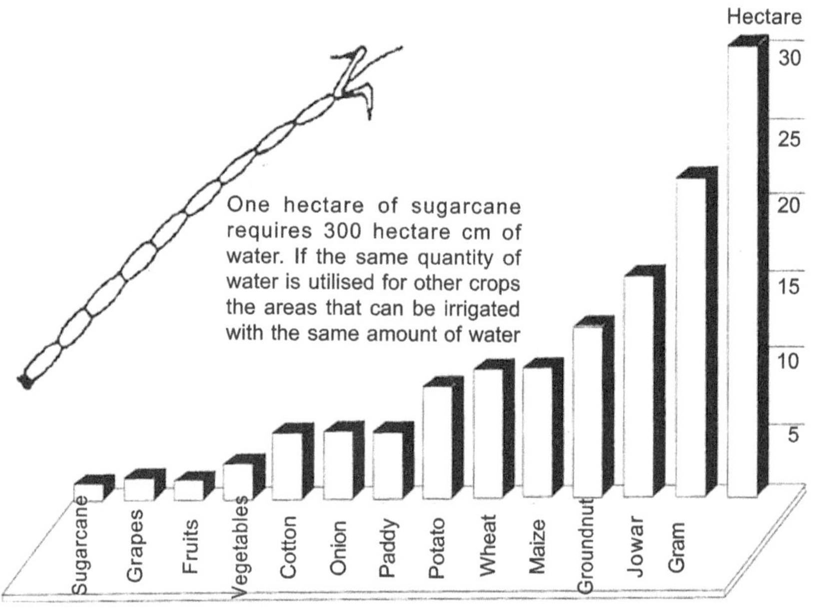

Fig. 13.5 Water needs of different crops
(Source : Falkenmark et al, 1990)

Since the water requirement of a crop is conditioned by rate of evapotranspiration in a particular situation, a knowledge of the rate of evapotranspiration (mm / d) is necessary to plan irrigation. The rate of evapotranspiration depends both upon meteorological factors (radiation, atmospheric humidity, temperature, wind, etc.) and field factors (wetness and surface properties). Potential Evapotranspiration (PET) corresponds to the "maximal evaporation rate which the atmosphere is capable of extracting from a well-watered field under given climatic conditions"

The crop water requirements (CWR) are computed from the following equation :

$$CWR = KC \times PET, \quad .....(13.1)$$

where KC is an empirical crop coefficient, and PET is Potential Evapotranspiration.

The KC values depend upon climate (higher values for hot, windy and dry climates, and lower values for cool, calm and humid climates) and the crop (stage of growth, crop reflectivity, crop height and roughness, degree of ground cover, canopy resistance to transpiration, etc.). For most crops, the KC value for the total growing season is in the range of 0.6 to 0.9.

It is important to bear in mind that soil water availability does not depend on the soil alone, but on the soil-crop-climate system. For instance, a crop with extensive and dense roots (such as those of small grains) can utilize the soil moisture more effectively than plants with sparse and shallow roots (such as those of potato). A crop growing in arid climates (with consequent high evaporative demand) tends to experience water stress more frequently than the same crop growing in moderate or humid climates.

Moisture conservation in drought-prone areas

Since the scarcity of water is the crux of drought, it is by taking techno-socio-economic measures to address the core issue that the vulnerability is reduced and the risk minimized. Thus in the drought-prone areas, the following techno-socio-economic measures could be taken to minimize the vulnerability and thereby minimize the risk. : (i) Conservation of water – through rainwater harvesting of rooftops, and surface runoff, groundwater recharge, (ii) Reserving some water sources, exclusively to be used for drinking water for humans and animals, (iii) Delinking of water rights from land rights – a farmer should have rights only for replenishable groundwater, and the deep groundwater is the heritage of the community as a whole, (iv) Only low-water need crops (e.g. millet or sorghum) should be allowed to be grown in drought-prone areas, and water is used efficiently through sprinkler and drip irrigation and use of low-cost green-houses

Techniques are available to improve the runoff efficiency, which is defined as the yield expressed as a per cent of the total precipitation on the catchment. Runoff efficiencies greater than 80 % can be achieved by covering the catchment with sprayed silicone compounds, concrete, aluminum foil, butyl rubber sheet, gravel-covered plastic sheet, paraffin wax, asphalt, polyvinylfluoride, etc. Apart from the problem of expense, there is an ever present risk of the materials being stolen. Besides, plastics are susceptible to damage by wind, sunlight, livestock or pests.

A labor-intensive, and inexpensive treatment is the method of soil grading and compaction. This does not require any purchased material, and it cannot be stolen. This can last for 5 - 10 years and costs only INR 2-3 (USD 0.04 - 0.06 cents) / m^2. The cost per 1000 L of yield is INR 3.50- 9.50 (USD 0.07 - 0.19 cents).

Irrigation Methods

The principal methods of irrigation, and their advantages and disadvantages are described as follows (Fig. 13.6; source: Hillel, 1987)

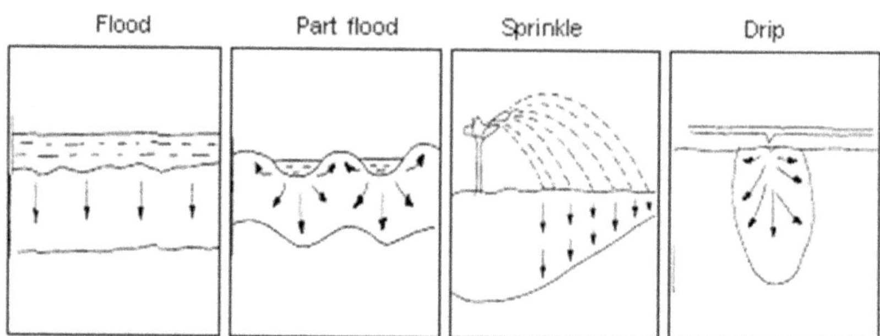

Fig. 13.6 Pattern of infiltration under different irrigation methods (Source : Hillel, 1987)

(i) *Surface irrigation or gravity irrigation*, whereby water introduced at the head of the field spreads and infiltrates throughout the field through forces of gravity and hydrostatic pressure. In this method, the soil is the medium through which water is *conveyed, distributed and infiltrated.* For this reason, the physical and chemical properties of the soil, and way the land has to be prepared for irrigation enter the picture prominently. This is the most ancient of all irrigation technologies, and it has been estimated that this method serves about 95% of the irrigated land worldwide. This method is mechanically simple, has low energy requirements (as it is based on gravity), and can be easily adapted to small holdings. Its serious deficiencies are low application efficiency, high conveyance losses and wastage of water, which lead to adverse consequences such as water-logging and salinization.

(ii) *Sprinkle irrigation*, whereby sprayed water falls on plants like rain. This system does not depend upon the soil surface for conveyance and distribution of water. It is hence not necessary to level the land (as required in the case of surface irrigation) and conveyance losses are minimal. Water can be applied at a rate less than that of the soil infiltrability. This allows soil aeration. The disadvantages are the high capital costs, maintenance requirements and energy costs (for maintaining high pressure). Also, sprinkle irrigation is strongly affected by wind.

(iii) *Drip irrigation*, whereby water is applied directly to the root zone through a set of polyethylene tubes laid along the ground or buried at a depth of 15 to 30 cm. Water is delivered drop by drop through perforations or emitters in the pipes. The trickling rate is maintained at less than the rate of infiltrability in the soil. The operating pressures are 15 psi (~104 kPa) to 45 psi (~310 kPa). Under this system, it is not necessary to level the land, and it is not affected by wind. By providing the rooting volume with the requisite amounts of water (and nutrients, which can be added to water), the soil can be kept continuously moist and aerated. As only a portion of the soil surface is wetted, the amount of direct evaporation is reduced. Slightly brackish water (upto 1000 mg/L) can be used under the method for some crops, which are salt-tolerant. As the water does not come into direct contact with foliage, it will not cause saline scorching. As the chosen volume of the soil is kept constantly wet, the salts in the brackish water do not have a chance of concentrating and affecting the crop.

Drip irrigation can be used any where in rugged terrains, in sandy soils of low moisture storage capacity, and arid climates of high evaporativity.

The capital costs of drip irrigation system are high. The method needs very rigorous adherence to scheduling and maintenance. In an effort to cope with the extreme scarcity of water, Israelis developed the drip irrigation into a fine art. In the context of rising cost and scarcity of water, and lower costs of plastic tubing, this method is likely to be widely adopted.

A microprocessor-based, drip irrigation system suitable for use in Developing countries, is now commercially available. This device uses low cost ceramic sensors, and operates on a solar cell-charged battery. The device continuously monitors the soil moisture and controls the drip rate so as to maintain the moisture within the desired limits.

(iv) *Micro-sprayer* : Micro-sprayers have several of the advantages of the drip systems in that the water is applied only to a fraction of the ground surface, high frequency irrigation is possible, and fertilizers can be injected, if desired. Besides, they have the following advantages over the drip systems : (a) As micro-sprayers have larger nozzle orifices, clogging of emitters is not a serious problem, and it is not necessary to filter the irrigation water, (b) Micro-sprayers are operated at pressures of the order of 2 atmospheres, which are much lower than for the drip systems, and (c) Micro-sprayers can be scaled down for use in small farms in the developing countries.

The micro-sprayers have the following minor disadvantages : (i) As wetting of leaves will be involved, brackish water cannot be used in irrigation, and (ii) Since the area wetted is larger area than in drip systems, there are some evaporational losses.

(v) *Low-head, bubbler irrigation* : A closed-conduit irrigation system has the great advantage in that it avoids conveyance losses and maintains uniformity in application. But any such system needs energy to pressurize water for distribution. Low-head, bubbler irrigation has the advantages of a piped system, but it does not require any pumps or nozzles. Even the low head available from a surface ditch may be adequate. In this arrangement, water is simply allowed to bubble out from open, vertical standpipes, 1-3 cm diam., which rise from buried lateral irrigation tubes. Bubbler systems are particularly suitable for widely spaced crops like fruit trees and grape vines. Small circular basins can be constructed around the trees for the water to bubble into. Low-volume, high-frequency, partial area irrigation is possible with the system. The initial cost of the system (about USD 1600/ha) is comparable or even lower than other systems. As the system works by gravity, there are no energy costs. It also has a longer life, because it is buried.

Methodology of optimum irrigation management

The problems in the case of irrigation arise from a fallacy in human thinking if some thing is good, more of it should be better. If a certain amount of irrigation raises crop productivity, which it surely does, more irrigation should produce more crops. As Hillel put it perceptively in his monograph, "The Efficient use of Water for Irrigation", *Just Enough is Best -no less and certainly no more.*

For too long, irrigation has been thought of as simply a water delivery system, and this had disastrous consequences. None would disagree with the dictum, *Just Enough is Best.* But to determine what is enough is an enormously complicated job.

The following scenario is envisaged for optimizing an irrigation system:

1. Determine the water requirements of the crop to be grown in a particular agro-climatic setting, on the basis of the Potential Evapotranspiration (PET) and empirical crop coefficient (KC) data,
2. Withdrawal of water from the source (river, reservoir or aquifer) in tune with the water requirements, on time-variable and space-variable basis,
3. Delivery of water in tune with the rooting depth of the crop, and available water capacity of the soil,
4. Planning the drainage system right at the outset as an integral part of the irrigation system. This is absolutely essential in the case of river valleys prone to high water-table conditions. Soil salinity should be monitored continuously in order to alert the farmer about the accumulation of injurious levels of salinity in the root zone of the plant.

   Advances in information technology have made it possible to optimize the various system variables. Efficiency of water delivery needs to be optimized in terms of conveyance of water with minimal losses (say, in closed conduits), capability to provide measured amounts of water calibrated to meet the needs of crops in time and space, while preventing wastage, salinity and rise in the water-table. Efficiency of water utilization is to be optimized to low-volume, low-pressure, high-frequency, partial-area irrigation to achieve high crop yields.

Planning of sustainable water supply systems

Sustainable development has been defined as that kind of development that "meets the needs of the present without compromising the ability of the future generations to meet their own needs". In the context of water resources, this could probably be worded as follows : Sustainable development of water resources encompasses those patterns of water resources utilization that will enhance social and economic benefits for the present and the future generations, without impairing the eco-hydrological processes. This implies that any use of water resources, which either leads to or has the potentiality to lead to, the degradation of the ecohydrological system is *ipso facto* unsustainable and hence unacceptable.

Schultz & Hornbogen (1995) point out that simply forecasting an attribute of water resources management, such as water demand, and designing projects to meet the future water demand, is inadequate. They propose an "anticipatory approach", whereby our predictions should involve not only parameters directly concerned with water, but also in terms all relevant variables, such as, hydrological and ecological conditions, as well as socio-economic conditions, including a potential change in the society's value system. Using the concepts of fuzzy-sets, and probability functions, they quantify a "possibility function" on the basis of a large number of possible future development paths.

Fig. 13.7 (source: Schultz & Hornbogen, 1995) illustrates the assessment of alternative strategies for possible future conditions in regard to water availability and water demand on the basis of the possibility functions for likely changes in hydrological and socio-economic conditions. By combining two one-dimensional possibility functions of (a) and (b), we can derive the two dimensional graph shown in (c), which takes care of the areas of low, medium and high water demand as well as water availability. Suppose the realization of sustainable development needs a new water project (say, a water supply dam) which is capable of addressing all possibilities. This would correspond to the outer edge of the rectangle specified as low possibility. Three curves are shown in (c) : performance of the present supply system, type A project alternative which covers the core conditions of high possibility, but which does not allow modifications, and type B which is close to the optimum solution but is flexible. If the project alternative B does not cost much more than A, it should be preferred as it allows greater flexibility. However, if alternative B is much more expensive than A, then one has to decide whether the extra expense is worth the flexibility.

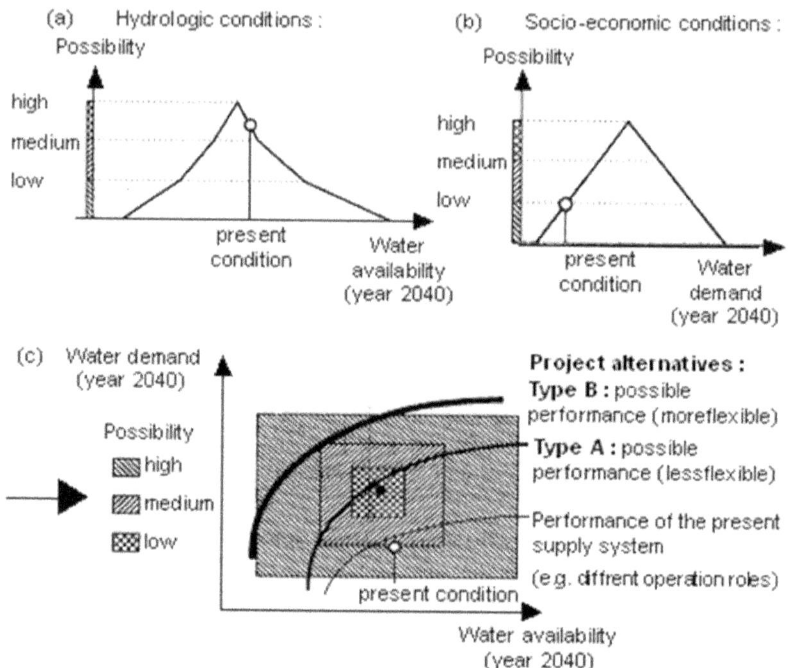

Fig. 13.7 Assessment of alternate strategies based on possibility functions (Source : Schultz & Hornbogen, 1995)

Decision-support system for water resources management

The water resource issues have so many dimensions and complexities that a decision-support system is necessary for water managers. Kaden et al. (1989) describe a simple computer-aided water management system, to assist operators, managers and planners (Fig. 13.7). It has three components:

(i) measuring and information systems (data acquisition, transmission and storage),

(ii) software system (user software),

(iii) Organizational system (organizational structure, legal and economic regulations).

Recommendations

There is little doubt that in the twenty-first century, water issues are going to emerge as the most serious problem facing humankind.

The visionary paradigm for water resources management that L'vovich proposed way back in 1979 continues to be valid to this day. It presupposes (i) cessation of the discharge of sewage into rivers and lakes, (ii) reuse of wastewater, (iii) closed circuit recycling of industrial water, (iv) underground reservoirs of the same order of storage capacity as surface reservoirs, (v) protection of forest cover to reduce runoff, and increase percolation and soil moisture, (vi) regulation of evaporation, (vii) water transfers, etc. (L'vovich, 1979, p.373).

Shiklomanov (1998) recommended the following steps to ensure the wise management of water resources :

1. Protection of water resources : Groundwater is sought to be protected through an understanding of its vulnerability.
2. Drastic decrease in specific water consumption, particularly in irrigation and industry : In future, the efficiency of irrigation has to be improved greatly (through measures such drip irrigation and cutting down of conveyance losses), so that with lesser total withdrawal (about 60 %), more food could be grown. Water consumption in industries is reduced by extensive recycling, and reuses various kinds of wastewater.
3. Complete cessation of the practice of discharging waste water into the hydrological systems (such as, rivers and lakes) : As pointed out by Shiklomanov (1998), every cubic metre of contaminated water discharged into rivers and lakes spoils up to 8 - 10 cubic metres of good water. This is a monstrous situation, and every effort should be made to stop this practice.

4. Harvesting of precipitation and making a more efficient use of runoff,
5. Greater use of salt and brackish waters : Saline agriculture using brackish water or sea water,
6. Influencing the precipitation-forming processes :
7. Use of water stored in lakes, underground aquifers and glaciers, and conjunctive use of water resources.
8. Spatial and temporal redistribution of water resources, through water transfers. :

References

Albinet, M. & J. Margat. 1970. Mapping of groundwater pollution vulnerability. *Bull. BRGM 2ème Series,* **3**(4), 13 - 22.

Aswathanarayana, U. (2001) *Water Resources Management and the Environment*, Lisse, Holland: A.A. Balkema.

Falkenmark, M., J. Lundquist & C. Widstrand. 1990. Water scarcity - An ultimate *constraint in Third World Development.* Tema V, Rapt. 14. Univ. of Linköping, Sweden.

Foster, S.S.D. & A.C. Skinner. 1995. Groundwater protection : the science and practice of land surface zoning. In *Groundwater quality : Remediation and Protection* (eds. Kovar, K. & J. Krásný). IAHS Publ. no. 225, p. 471- 482.

Van der Heijde, P.K.M. (1992) Computer modeling in groundwater protection and remediation. In P. Melli & P. Zannetti (eds.) *Environmental Modeling*: 1-21, Amsterdam: :Elsevier

Hillel, D. (1987) The efficient use of water in irrigation. World Bank Tech. Paper no. 64, Washington, D.C. : The World Bank.

Kaden, S., A. Becker & A. Gnauck. 1989. Decision-support system for water management. In *Closing the Gap between Theory and Practice* (eds. Loucks, D.P. & U. Shamir), IAHS Publ. no. 180, p. 11 - 21.

L'vovich, M.L. 1979. *World Water Resources and their future.* Washington, D.C. : Am. Geophy.Uni.

Meybeck, M. (1998) Surface water quality: Global assessment and perspectives. In H. Zebidi (ed.), *Water: A looming crisis*? Paris: Unesco, p. 173-185.

Schultz, G.A. & M. Hornbogen. 1995. Sustainable development of water resources systems with regard to long-term changes of design variables. In *Modeling and Management of sustainable basin-scale Water Resources Systems* (ed. Simonovic, S.P. et al.), IAHS Publ. no. 231, p. 329-338.

Shiklomanov, I. 1998. *World Water Resources: A new appraisal and assessment for the 21st. century*. Paris : UNESCO.

UNEP (1983) Rain and Storm Water Harvesting in Rural Areas. Dublin: Tycooly.

Wang, B. (!991) Generating and resources recoverable technology for water pollution control in China. In B.Z. Wang et al (eds.) Low-cost, and energy-saving waste water treatment technologies. *Water Sci. & Tech.*, 24(5), 9-19.

UNIT - III

Soil Resources

14

In situ Conservation of Rainwater on Alfisols in a Semiarid Region

Prabhu Prasadini

Dept. of Environmental Science & Technology, College of Agriculture
ANGRAU, Rajendranagar, Hyderabad – 500 030, INDIA.
(e-mail : prabhuprasadini@rediffmail.com)

Key words : in situ rainwater conservation. Alfisols. Contour cultivation. Clay Mixing. Crop residue incorporation. Enhanced crop output

The twin problems constraining agricultural production are firstly, the all pervasive degradation of natural resources such as land, water and biodiversity, and secondly, fast declining soil fertility and low efficiency inn the utilization of inputs such as water, fertilizer and energy. Demographic pressures accentuate the first problem, and agronomic deficiencies, the second problem.

Alfisols have Al and Fe in the B-horizon, and hence the name. They are moderately leached soils that have relatively high native fertility.

The productivity of alfisols is critically dependent upon the adequate availability of soil moisture. The chapter deals with ways and means of conserving the soil moisture in alfisols to optimize their productivity.

In India only about 20 per cent of the irrigated land produce more than one crop a year and, as shown by the data of Table 14.1, yields of irrigated crops were just about twice those of rainfed crops. Table 14.1 also shows that with improved dry land practices, yields on rainfed lands could be equal to those of irrigated crops (Bentley, 1979).

Table 14.1 : Comparative yield indices for irrigated and non-irrigated crops, and comparative yields with improved and traditional practices, in semi-arid regions of India

Crops	Yield indices			Non-irrigated yields, q/ha		
	1	2	3	4	5	6
	Irrigated	Non-irrigated	(1) / (2)	Improved practices	Traditio-nal practices	(4) / (5)
Barley	230	100	2.3	29.3	12.8	2.3
Maize	390	100	3.9	28.0	12.2	2.3
Sorghum	490	100	4.9	25.2	5.2	4.8
Wheat	170	100	1.7	24.1	8.5	2.8
Pulses	100	100	1.0	17.7	5.5	3.2

[1]Yields for improved practices are each averages for at least 4 to 6 locations and 3 to 4 seasons at each location.

[2]These yields for traditional practices are close to well established average yields for the districts concerned.

The potential productivity of rainfed agricultural areas depends on the amount and distribution of rainfall, favorable temperatures for growth of crops, inherent fertility of soils, good tilth or physical condition, besides the absence of harmful, competing, or destructive factors such as salinity or soil acidity, weeds, insects, and diseases.

Agriculture is the major sector which utilizes as high as 85% of the available water in the country. However, the availability of water to this sector is going to be reduced to 68% by 2050, because of more demand form domestic and other purposes. Apart from quantity, the quality of water is also deteriorating because of the increased human activity. Apart from water, the other important natural resource essential for agriculture is land. Its per capita availability is also decreasing and is estimated to be 0.08 ha only by 2020 in our country because of galloping population growth. Under this situation utmost care has to be taken to use them in a most sustainable manner to meet the needs of the population with time.

In Andhra Pradesh nearly 65% of the area is under rainfed agriculture. The vagaries of monsoon decide the fate of our agriculture output. The commonest limitation on agricultural production is the shortness of the rainy season, its variability from year to year, its erratic monthly distribution and the very considerable moisture looses by runoff during intense rains. In the Hyderabad (study) area the rainy season is less than four months. Eighty percent of the precipitation comes in about 110 days usually between June 25 and October 10. There was 1400 mm of rainfall in the wettest year, and less than one-quarter of that amount (only 396 mm) in the driest year (1972). The average annual rainfall is about 725 mm, but the coefficient of variation is over 25 percent.

The crop yields in rainfed area depend on the onset, quality, distribution and withdrawal of the monsoon. In Andhra Pradesh average rainfall varies from 300mm in southern agroclimatic zone to >1000mm in High Altitude zone, being 700-900mm in the study area. Out of an average of 896 mm of rainfall in our state, nearly 67% is received during south west monsoon, 22% during north east monsoon, 2% in winter and 8% in summer. Duration of the south west monsoon is short (June-September) *i.e.,* less than 4 months. Efficient management of rain water on watershed basis is crucial for the improvement of rainfed areas. The soils in rainfed areas are not only thirsty but also hungry. Research data indicated that as high as 40%

of rain water goes as runoff causing soil and nutrient loss, 41% as evapotranspiration, 9% gets percolated to ground water bodies and only 10% gets retained as soil moisture. Apart from loss of rain water, nearly 75 to 100 tonnes of valuable top soil is lost through erosion in about 8 million hectares in Andhra Pradesh. For the sustainable agriculture soil and rain water management, which always go together is a must in rainfed areas.

The rain water can be used effectively by reducing the runoff losses (through *in situ* conservation, water harvesting and reuse), selection of appropriate crops and cropping systems (cover, phenology, organic matter addition) and soil management practices (through increasing surface roughness, increasing porosity, improving soil structure).

The *in situ* water conservation is more feasible and practical proposition under most situations. Hence the studies were made in the study area where the soils were light textured and of hard setting nature because of low clay content and low organic matter with 2-3% slope, with underlined *murram*. The *in situ* conservation practices can broadly be grouped into agronomic practices, like summer ploughing, contour cultivation, contour ridging, conservation furrows, deep ploughing, vegetative barriers (Tables 14.2, 14.3, 14.4 & 14.5): clay mixing (Fig. 14.1), crop residue incorporation (Fig. 14.2), suitable crop and variety, intercropping, strip cropping *etc*. and engineering practices, like land leveling, bunding (stone/gravel, contour bunds, graded bunds), terraces, contour trenches *etc* (Table 14.6).

The excess rain water over and above the intake rate of soils which will go as runoff is to be harvested and channeled through grassed water ways safely to the different storage structures, like farm ponds, check dams, percolation tanks, small and medium tanks etc. The harvested water should be used for life saving irrigation and high value crops.

Conservation tillage in the study area

As the soils are sloping with 2-3% slope and underlined with *murram* (soft gravelly admixed with clay, high bulk density layer) cutting the slope effect through contour cultivation, vegetative barriers, contour ridges, conservation furrows and breaking the sub-soil *murram* layer through deep tillage using sub-soiler and their combinations were tried (Singa Rao *et a.l*, 1995 and Prabhu Prasadini *et al.,*2001).

Contour ridges and furrows facilitated in collection of rainwater in the furrows to a great extent resulting in reduced runoff. Since the runoff was reduced, the soil, which would be carried away by the running water, was also reduced (Table 14.2).

Table 14.2 : Effect of *in situ* conservation practices on run off soil loss and castor yield (1993-94)

Treatment	Runoff (mm)	Soil loss (t/ha)	Castor Yield (q/ha)
Contour cultivation	30.8	2.54	8.14
Vegetative barrier	10.55	0.62	8.79
Conservation furrow	16.44	1.22	8.73
Conservation furrow + vegetative barrier	7.18	0.36	9.23
Ridges and furrows + vegetative barrier	4.05	0.24	9.75
Sowing along the slope	35.14	2.98	7.47
Rinfall in the crop period (mm)	234.1		

There was more infiltration time and more water entered the soil resulting in higher soil moisture (Table 14.3). Conservation furrows formed on contours at a distance of 3.6 m acted as temporary reservoirs of rainwater that was collected from the 3.6 m land above the furrows and retained the water till they were full. So the runoff was reduced and hence, the soil loss was also low (Table 14.2). Deep tillage loosened the soil and thus facilitated in more infiltration of rainwater in to the soil resulting in higher moisture (Table 14.3). As the clay admixed with murram also contributes to water storage, deep tillage is suggested to facilitate root penetration by loosening the sub-surface *murram* layer. Care should be taken to avoid inversion i.e., preventing the sub surface *murram* to get mixed with the surface soil. That's why sub-soiler which opens or just breaks the sub-surface murram is suggested for deep tillage. Vegetative (*Khus*) barrier acted as a barrier and slowed down the velocity of running water, thus giving more opportunity time for infiltration. With higher soil moisture regime in the conservation treatments (Table 14.3), the yield of crops was also higher (Table 14.4).

The extra yield the farmer could obtain with improved practices is also shown in Table 14.5.

Table 14.3 : Effect of conservation practices on soil moisture (% w/w) in 0-45 cm depth over growth period

Treatment	1997-98	1998-99
% increase over sowing along the slope		
Contour cultivation	6.1	8.1
Vegetative barrier	9.8	11.3
Conservation furrow	9.1	11.2
Conservation furrow +	16.0	18.2
vegetative barrier		
Ridges and furrows	19.9	23.7
Ridges and furrows +	23.7	28.2
vegetative barrier		
Deep tillage	26.6	22.9
Deep tillage + vegetative barrier	28.9	25.9
Rinfall in the crop period (mm)	307.4	931.3

Table 14.4 : Effect of *in situ* conservation practices on rainfed crops yield

Treatment	Yield (% increase over control)			
	1996-97	1997-98	1998-99 (Intercrop 2:1)	
	castor	Sorghum	Sorghum	Pigeonpea
T1 - Contour cultivation	15.1	16.4	16	14.2
T2 - T1+ vegetative barrier	19.9	23.9	20.8	20.2
T3 - Conservation furrow	20	23.3	21	20.9
T4 - T3 + Vegetative barrier	26.9	28.8	26.8	25.9
T5 - Ridges and Furrows	30.1	33.6	31.1	29.4
T6 - T5 + vegetative barrier	35.9	39	35.2	34.3
T7 - Deep tillage	29.9	41.1	32.1	32.0
T8 - Deep tillage + Vegetative barrier	33	50.7	33.9	33.0
Rainfall during the crop season (mm)	726.9	307.4	931.3 (no rain after harvest of Sorghum)	

• control – sowing along the slope

Table 14.5 : Extra yield per unit quantity of extra water conserved (kg/ha/mm) through soil management

Treatment	1993-94	1994-95	1996-97	1997-98	1998-99
Sowing along the slope	-	-	-	-	-
T1 - Contour cultivation	15.35	25.4	12.3	88.8	41.5
T2 - T1+ vegetative barrier	6.31	10.1	3.5	29.3	12.6
T3 - Conservation furrow	6.72	13.1	4	33.1	13.9
T4 - T3 + Vegetative barrier	6.20	12.0	4.1	30.0	13.6
T5 - Ridges and Furrows	-	-	4.9	37.9	16.8
T6 - T5 + vegetative barrier	7.2	14.2	4.9	36.8	16.1
T7 - Deep tillage	-	-	14.8	110.5	45.6
T8 - Deep tillage + Vegetative barrier	-	-	10.24	41.2	21.4

Vegetative *(Khus)* barrier

The runoff losses of the rainwater were reduced by about 60% there by the associated soil loss was reduced by about 65% and the soil moisture content available for the crop was increased. The average soil moisture content during the crop growth period of rainfed crops was recorded higher when the vegetative barrier was planted on contours. On an average over the four years, it was 12.54; 13.61 and 15.97 per cent higher in 0-15, 15-30 and 30-45 soil depths, respectively over the soil moisture content where the *khus* barrier was not taken up. The cumulative beneficial effect was reflected in rainfed crops yields. The per cent increase in rainfed crops yield was 22.1 (jowar), 2.13 (red gram) and 22.0 (castor).

Land management practices

Effect of some of the land management practices on reduction of rainwater runoff losses and the resultant of that *i.e.,* increase in the rainfed crops yield in the region of the study is shown in Table 14.6 (Padma Raju et al, 2002).

Table 14.6 : Effect of land management on runoff and crop yields

Land management	Runoff Reduction (%)	Crop	Yield increase(%)
Deep tillage	4646	Sorghum	31
Scoops(40mm)	6969	Sorghum	28
Tied ridges	39	Maize	35
FYM[a] (15 t/ha)	51	Maize	56
Straw mulch[a] (15 t/ha)	76	Maize	53
Prior perennial[b]	79	Maize	36
Contour bunds[c]	37	Maize	NA
Graded bunds[c]	31	Maize	NA
Porous barrier[d]	66	Maize	NA
Sand mulch[e]	80	-	-

a. = Effect after 4 years of continuous application

b. = Effects are during the 1st year after removing a 4 year mixture of perennial pigeonpea, Cenchrus and stylo

c. = average of 4 years

d. = porous barriers are 3 years old

= average of 2 years after bringing the sand cover of the surface to 95%.

Clay mixing technology

In A.P., about 40% of the total cultivated area is occupied by light textured red sandy loam and loamy sand type of soils. The clay content of these soils is low (<15 per cent) with low water holding capacity (5-10 cm m^{-1} depth) and is susceptible to leaching losses. As these soils are mostly (around 80 per cent) under rainfed cultivation, their low water storage capacity of the rain water is a major constraint. Studies were made to reduce the percolation losses of rain water and improve water holding capacity by mixing the locally available heavy textured soil in surface 0-15 cm soil.

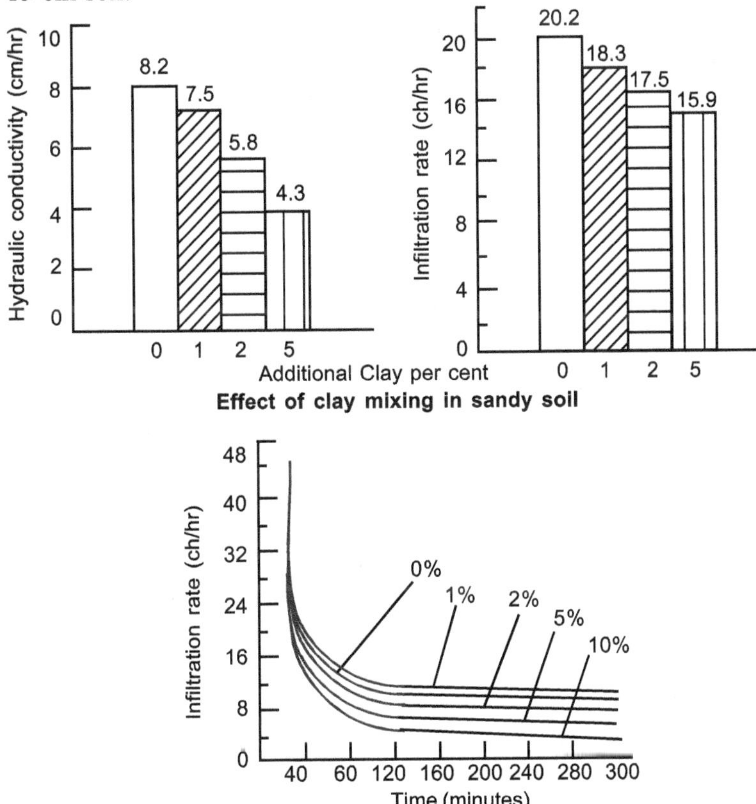

Fig. 14.1 Benefits of clay mixing on soil conductivity parameters

Considering operational feasibility and economics, addition of heavy textured soil to increase clay content by 2 per cent at a time, for a period of 2 to 3 years is recommended to farmers by ANGRAU. On farm trials were carried out in the study area with 2 per cent clay addition technology. Clay mixing modified the soil physical environment in terms of bulk density, hydraulic conductivity, infiltration and water retention. It decreased bulk by about 0.03 to 0.07 g cm^{-3} (2 to 4 per cent) and increased soil moisture content by 6.5 to 23.5 per cent. Hydraulic conductivity was decreased by 32 and 40 per cent and infiltration rate was decreased by 60 and 40 per cent in soils with 5 and 12.5 per cent initial clay, respectively. Thus, the improved soil physical conditions resulted in higher yields of sorghum, tomato and *bhendi* grown under rainfed conditions by 16.0, 10.8 and 10.5 per cent, respectively. The benefit cost ratio worked out was 1.37 with vegetable crop in the study area in one season. This beneficial effect of clay addition is permanent as it modifies the soil texture and particle linkages and thereby soil physical environment and ultimately yields of rainfed crops every season.

Crop residue incorporation technology

Red sandy loam soils (chalka) which cover a large area in A.P., become very hard on drying. As a result, rainwater is lost as runoff and growth of crops is adversely affected in rainfed areas. Incorporation in these soils of slowly decomposing crop residuals, such as, paddy husk, and powdered groundnut shells, will have the effect of keeping the soil porous and increasing the infiltration rate of the rain water. The soils of the study area are also of the same type, and hence this technology was tried. In the farmers' fields, the influence of incorporation of paddy husk @ 50qha^{-1} in the red sandy loam soil and its residual effect upto 2 years on soil properties and crop yields, was studied. During first year of incorporation, sorghum (CSH-5) was taken as test crop and residual effect was studied with castor (2nd year) and sorghum (3rd year). The incorporation of paddy husk @ 50 q ha^{-1} in red sandy loam reduced slightly the bulk density of 0-15 and 15-30 cm layers, increased the amount of moisture retained in these layers by 1.8 per cent and 2 per cent respectively, (Fig. 14.2).

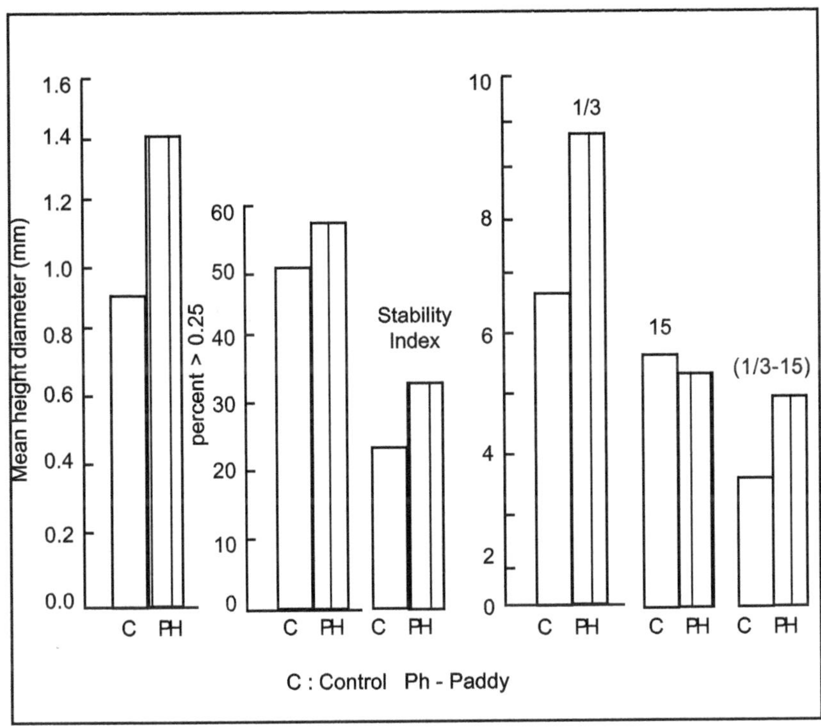

Fig. 14.2 Effect of incorporation of paddy husk on aggregation and soil moisture retention

The effect of incorporation of paddy husk persisted upto the harvest of crop. There has been Improvement in the structural indices, namely, mean weight diameter, percentage of aggregates greater than 0.25 mm diameter and aggregate stability of soil in 0-15 cm layer. The amount of moisture retained in 0-15 cm layer at 0.33 MPa increased slightly. However, it decreased slightly at 1.5 MPa and available moisture (0.033-1.5 MPa) increased with the incorporation of paddy husk. The final infiltration rate of red sandy loam increased from 8.2 to 11.0 cm ha^{-1} and the cumulative infiltration from 45.9 to 57.7 cm with the incorporation of paddy husk. This improvement in soil structure may remain for longer period as per cent organic carbon content was increased significantly by 0.06 with one incorporation. Yields of rainfed sorghum (1st yrs), castor (2nd year) and sorghum (3rd year) increased by 33.08, 22.99 and 13.56 per cent with the incorporation of paddy husk, indicating its residual effect. The benefit:cost ratio was 3.09. Repeated photographs of the study area clearly demonstrate how a waste land could be converted to good crop yielding land by adopting the above mentioned soil management practices.

The conserved water both in fields and water bodies is considered as "liquid gold". Hence in dry land agriculture care has to be taken to limit possible losses such as evaporation, deep percolation and seepage, through the use of procedures such as, mulching (soil, organic and plastic), anti-evaporants, lining (clay, cement and plastic) and appropriate method of irrigation (drip, sprinkle) etc. Other improved technologies like weed management, fertilizer management, INM, IPN, cultivation if improved genotypes should be adopted to the maximum possible extent as these are complementary to the water management practices in improving the water use efficiency.

References

Bentley, C.F. (1979) Agriculture potentials awaiting exploitation. Jannareddy Venkat Reddy Memorial Lecture 1979, ANGRAU.

Padma Raju, A., Singa Rao, M. and Prabhu Prasadini, P. (2002) Technologies for efficient water management in agriculture. Workshop on the "The water crisis – hope and action for humanity's future" on 29th November, 2002.at National Geo Physical Research Institute, Hyderabad, INDIA.

Singa Rao, M. ,Prabhu Prasadini, P, Nagender Rao,C and Jayasree G. (1995) Soil physical constraints and their management for increasing crop production in Andhra Pradesh, AICRP on Soil physics, ANGRAU.

Prabhu Prasadini, P., Nagendar Rao.C. and Jayasree.G. (2001). Resource characterization and management and soils of Andhra Pradesh. AICRP on Soil physics, ANGRAU.

UNIT - IV

Mineral Resources

15

Mineral Resources Management Methodologies

U. Aswathanarayana

Mahadevan International Centre for Water Resources Management, Hyderabad, India.

(e-mail : uaswathanarayana@yahoo.com)

Key words : Adverse impacts of mining. Environmental sustainability. Artisanal mining. Mined land. Beneficial use. Mining wastes. Mine water. Gold processing technologies.

"I must be good for some thing. God makes no junk" –
An American school boy poster

Introduction

This chapter is focused on employment-generating activities involved in the reduction of the adverse impact of mining, rehabilitation of mined land, and the reuse of mine wastes and mine water, etc. Readers interested in issues such as Mining methods and environment, Process technologies in relation to environment, Control technologies for the minimization of the environmental impact of mining, and ways and means of mitigating the mining impact, etc. summarised by Aswathanarayana (2003, "*Mineral Resources Management and the Environment*", A.A. Balkema)

Mining, like the proverbial serpent in the Garden of Eden, has never been held in high esteem. Most people consider mining as an unmitigated evil, and some who are more realistic, concede that it is a necessary evil (but evil all the same). This is not a new development. In the olden days, mines were invariably worked by slaves - chained to pillars underground, the slaves used to die in a matter of weeks. In the medieval Europe, being condemned to work in the salt mines was a form of punishment worse than death.

Presently, the horrendous consequences of mining are evident everywhere. The landscape in some countries (e.g. USA, Zambia, PNG) is pockmarked with gigantic pits. As pointed out by F¯rstner, the mass of the mine tailings produced worldwide (18 billion m^3/ year) is of the same order as the quantity of sediment discharge into the oceans. As progressively lower grades are worked, the mass of the mine tailings is expected to double in the next 20 – 30 years. Vast areas are either strewn with rock fragments, and in some areas, Acid Mine Drainage has rendered the soil and water so acidic that not a blade of grass grows there. Whole towns (e.g. eastern India) had to be abandoned due to subsidence caused by underground coal mining. Mine workers are exposed to a number of physical, chemical, biological and mental hazards, and mining is ranked as number one among the industries in the average annual rate of traumatic fatalities.

This high-tech model used by the Industrialized countries to minimize the number of workers and increase the output, is, however, not applicable to the Developing countries, for the following reasons: (i) the investments needed are high – for instance, a block cave mine may need an investment any where from USD 100 million to 1000 million, besides requiring very highly skilled personnel, (ii) what the developing countries need is *job-led* (and not job-less) economic growth. A sensible strategy for the developing countries is to use the mining industry to promote *job-led* economic growth through the adoption of employment-generating, economically-viable and environmentally-acceptable technologies.

Environmentally-sustainable mining

The vision of UNEP (Hoskin, 2001, Proc. Int. Conf. on *Mining and the Environment*, Skellefteå, Sweden) to achieve environmentally-sustainable mining industry in the early part of the twenty-first century, has the following components:

1. Recycling of metals should approach 100 % - recycling of metals reduces disposal pressures, and results in great energy savings. The limited amount of virgin metal that may be needed should be obtained from highest-grade reserves.
2. *Technological improvements :* (i) Application of remote sensing and hydrospectral analysis in exploration, and monitoring of tailing impoundments, closed mines, and compliance with environmental regulations, (ii) minimization of mine wastes, reducing air and water pollution to essentially zero level, and fabrication of lighter, stronger and more durable materials, and secondary recovery of useful materials from mine wastes, (iii) Remediation of abandoned mine sites to increase arable land for agricultural production.
3. One of the major sources of accidents in the mining industry is the failure of tailings' dams. This problem can be mitigated by (i) new technologies to dewater waste slurries – production of paste-consistency material from mill tailings, (ii) A thorough analysis of all the design components including site selection, drainage systems, impoundments, measurements and inspections needed with respect to water balance, taking into consideration unusual conditions arising from rain, ice and snow and seismic activity.

4. Training and assistance to small-scale miners, particularly in Africa, Asia and Latin America, to improve the commodity recovery, reduce environmental damage, and improve local health and safety conditions. Phasing out of the use of mercury in artisanal gold mining.

Struthers (2001, Proc. Int. Conf. on *Mining and the Environment*, Skellefteå, Sweden) proposed is a new holistic Total Project Development (TPD) approach to mining. Under this approach, a mining project is developed as a part of much wider, multi-activity regional development. All the material extracted by a mining company is put to productive use. Waste rock, mine tailings, excess mine water, etc. are used as raw material for a variety of downstream ancillary industries. Tailings are used for underground backfill, embankments and sealants for reactive waste rock, and production of construction materials for mine use. All excess tailings are used for "soil" development. Excess process water (after use in recycling) is treated for being used in fish farm ponds and crop irrigation

The TPD approach benefits the various entities in the following ways:

Mining company benefits from (i) increased metal recoveries and additional revenue through the retrieval and sale of non-target minerals, (ii) reduced operational expenses through the maximum utilization of tailings in backfill, preparation of tailings/concrete blocks for underground and surface constructions, and replacement of topsoil, (iii) reduced mining costs by saving on the construction of tailing ponds and waste rock dumps, (iv) reduced rehabilitation costs, and (v) income from productive use of post-mine land use. The mining company would also have some intangible benefits such as good public image, making it easier to get the required environmental permits.

The local community benefits from increased employment and income levels, long-term food security and livelihood after the closure of the mine. The improved infrastructure helps in communications, and access to wider markets. TDP benefits the environment through the elimination of waste rock dumps, drastic reduction in Acid Mine Drainage (AMD) and contamination from dust, heavy metal contamination of soil and water, etc.

Iron ore mine, Kiruna, Sweden – A case study

The LKAB iron ore mines in Kiruna, Sweden, are located above the Arctic Circle. Two underground mines, located at Kiruna and Malmsberget, employ sublevel caving method, to produce 30 Mt / y of iron ore. Much of the mining operation is automated – for instance, the production drilling rigs, loading machines and transport systems on the new main haulage level are remote-controlled. The magnetite ore is crushed, ground, screened and upgraded by magnetic separation techniques. The iron ore concentrate is pelletised using bentonite as a binder.

The LKAB complex demonstrates how continuous improvements in technologies could bring about high productivity, while at the same time reducing water and air pollution (Nordstrom, Proc. Int. Conf. on *Mining and the Environment*, Skellefteå, Sweden 2001, p. 604-614). Mining and process water goes through large pond systems, to facilitate the removal of the sludge by sedimentation. The water is then pumped to the clarifying ponds from which the process water is recirculated. Since 1980, LKAB's atmospheric emissions have been halved, while the production of pellets has doubled. The external energy consumption in the pelletising plants which was 639 MJ (million joules) / t of pellets in 1970 has been brought down to 250 MJ/t in 2000.

The following air pollutants are produced in the process of pelletisation:

(i) Fine dust, mainly composed of iron oxides, before induration,

(ii) Sulphur dioxide, from sulphur in the fuel, and sulphides in green pellets,

(iii) Hydrogen fluoride and hydrochloric acid from apatite residues in green pellets,

(iv) Nitrogen oxides, from nitrogen in the fuel and in the atmosphere.

That the emissions in the pelletising plants in Kiruna are less than the statutory limits is evident from the following information :

Parameter	Statutory limit (g/t of pellets)	Emissions in Kiruna (g/t of pellets)
Particles	100	60
Sulphur dioxide	15	13
Hydrogen fluoride	6	2
Hydrochloric acid	6	2

Noise is generated by ventilation fans, mining and transport equipment. The maximum noise level nearest to the dwelling should not exceed 40 - 45 dB (A). Blasting in mines produces vibrations in the bedrock in the surrounding areas. Noise and vibrations are measured continuously to minimize the inconvenience to property owners in the neighbourhood.

Artisanal Mining

The contribution of small-scale or artisanal mining to the overall mining output in the world is estimated to be 10-16%. In general, the percentage contribution with respect to industrial minerals is higher than for metallic minerals. Estimates vary widely about the number of persons involved in artisanal mining (upto 16 million). In some countries, the contribution of artisanal mining is of considerable economic significance. For instance, Peru has about 3,000 small-scale mines which produce 100% of the antimony, 90% gold, 15% tungsten, etc. of the national production (Gocht, 1980, *Natural Resources and Development*, v.12, pp. 7-18). About 90% of Brazil's gold production is attributable to artisanal mining.

Capital-intensive, mechanized mining is not cost effective for the exploitation of small deposits, even of high grade. Such a deposit is amenable to small-scale mining, particularly where the deposit occurs at or close to the surface, and where the mineral could be won by simple methods, such as hand-picking, panning, sluicing, etc. It therefore follows that artisanal mining may be a cost-effective option in the case of several economic minerals occurring in the soil. For instance, by the nature of its occurrence and its properties, opal can only be mined manually.

Small-scale mining has several advantages: it is labor intensive, can be initiated on any scale, with simple technology, at low capital cost, low consumption of energy, and short lead time, and without expensive imported equipment. It can also promote local industries. This, however, does not mean that small-scale mining is the panacea for the problems of the developing countries. Artisanal mining suffers from the following serious disadvantages: (i) it tends to be haphazard, since in most countries there is no systematic exploration activity to support small-scale mining; (ii) destructive exploitation by the gouging of rich pockets; (iii) low recovery rates; (iv) low labor productivity (about 4% of highly mechanized mines); and (v) non-extraction of valuable by products which are therefore irretrievably lost to the country.

In future, vegetative methods of reclamation of mined land may emerge as a significant, economically viable, and employment-generating activity.

Innovative technologies suggested

It is possible to improve the efficiency of small-scale mining, while concomitantly reducing its deleterious consequences, by the adoption of the following innovative approaches :

1. Developing simple techniques of prospecting, which could be used by unskilled labor, e.g. use of smoky quartz as indicator of cassiterite-lepidolite pegmatites, and looking for cassiterite resistate in the soils near pegmatite. Training of miners on-site about simple methods of mineral search and extraction.

   Using a portable X-ray fluorescence analyzer, it is possible to make a quick and fairly accurate on-site assay of several ore metals in the material mined or to be mined by a miner. Such an assay can serve two purposes : (i) to make the miner aware of the economic value of the material already mined by him (through a knowledge of what kind of ore metals and in what concentrations occur in the material mined by him), and (ii) to advise him as to what kind of material he should be mining in order to get greater returns.

2. Research and development to design improved methods of ore search and ore extraction relevant to small-scale mining. Placer gold is a case in point. An artisanal miner can extract gold only if it is coarse grained (say, 30 μm) and high grade (say, about 25 g /m^3). He uses the mercury amalgam method of extraction which is highly polluting. New carbon-in-pulp and carbon-in-leach technologies have several advantages : (i) they are capable of extracting fine-grained gold (about 10 μm) and at low concentrations (about 2 g/ t); (ii) they are environmentally benign. These technologies need to be adapted for small-scale operations. In extremely dry areas, pneumatic methods of gold separation have to be developed.

3. Using mobile units for preconcentration and extraction on site: Truck-mounted, diesel-powered, self-contained, ore-dressing modules are taken to the site of the artisanal mining and the ore is concentrated / extracted on site. The mobile unit can be owned and operated by a cooperative or a private company. A part (say, one-third) of the output could be collected in kind towards service charge due to the mobile unit and the royalty due to the government. As the recovery rates by the mobile unit are at least 2 - 3 times higher than by manual methods, the artisanal miner is still left with considerably more saleable material than he would have been able to recover on his own.

4. Through the use of the mercury amalgam method, artisanal gold mining industry in Tanzania has severely contaminated the waters and soils. About 78% of the water samples analyzed contained concentrations of mercury higher than the permissible level of 1lg/ 1 . Mercury levels in the mine tailings range from 1.31 to 18.7 lg/g. The Institute of Production Innovation of the University of Dar es Salaam, Tanzania, has developed a rugged, easy to handle, highly portable and locally manufactured, and inexpensive (eq. USD 50) retort which has efficiency of 99.6 to 100 %. The use of such a retort is the most effective way to reduce the pollution of airborne mercury produced by the firing of the Au – Hg amalgam..

5. The "Portable" gold plant developed by Libenberg, Rundle and Storey of San Martin mining company (*Mining Magazine*, July, 97, pp. 8-10) is a veritable godsend for small-scale gold miners. The salient points of the plant are as follows: San Martin 's claims encompasses two dumps around Bonda, Kenya, with 250,000 t of material, grade : 1 - 3 g / t. Carbon-in-pulp / carbon-in-leach technique; Capacity of the plant: 10,000 t / month; production cost: USD 150 / oz. The total steel requirement (12 t) for tanks, baffles, agitator mountings, and the pumps and piping, were brought from South Africa in one container and erected on site. Dump material is reclaimed by high-pressure water. In the first leach tank, lime (5 kg / t) and sodium cyanide (0.5 kg/ t) are added. The residence time in each of the absorption tanks is approximately 1.3 hr at a throughput of 10,000 t / month, and carbon concentration of 15 - 20 g / l. The eluate is heated in a diesel-fired cast iron burner. Gold is recovered onto steel wool cathodes. Security is ensured by having the recovery cell protected by a 220 V inner cage, and 15,000 V outer cage (powered by solar cells). Doré is 80% pure. Total power consumption: 145 kw (diesel generators). Water is pumped from the Yala River (3 km from the plant).

   The plant has been in operation for several years now and has the following advantages: environmentally-benign as no mercury is used; can be erected even in remote areas, and shifted and reassembled without much problem; can be operated with minimal expatriate assistance; economically viable. This technology can be used in two kinds of situations : (i) as a private enterprise, for treating dumps, where they exist, and (ii) as a cooperative, by setting up the plant at a central place where a number of artisanal miners (50 to 100) operate.

6. The modular plants designed by M/S Bateman Project Holdings Limited, Boksburg, South Africa, have revolutionized the recovery efficiencies in artisanal mining of a number of minerals (such as, modular process plants for prospecting, sampling and small scale production of diamonds). The heart of the equipment is the Dense Media Separator (DMS) plant, which uses water-ferrosilicon mixture as the flotation medium. The density of the medium is automatically controlled. The smallest unit has a capacity of 1 t/h of diamondiferous gravel in the size range of 1 mm to 8 mm. The modular unit which has become very popular is 5 t/h DMS unit, which weighs only 4.5 t, and can fit into a standard 12 mm container, with all the spares. It is custom-designed depending upon the kind of feed, assembled and tested. It is then dismantled and shipped to site. It can be erected on site in a day or two. The DMS modules are skid mounted and has built-in spillage pumps, and no concrete foundations are needed.

Rehabilitation of Mined Land

Though the mining companies are required by law to submit plans and commit funds for the rehabilitation of the mined land when once the mine is closed, enforcement has not always been strict enough. It is particularly difficult in the case of artisanal miners ("here today, gone tomorrow").

Though the climate, soil and hydrological characteristics and methods of mining vary greatly in different areas, there are some common elements in the techniques of rehabilitation (i) Removal and retention of top soil, to be respread in the area that is being rehabilitated, (ii) Reshaping the degraded areas and waste dumps in such a manner that they are stable, well drained, and suitably landscaped for the desired long-term use, (iii) Minimizing the potentiality for wind and water erosion, (iv) Deep ripping of the compacted surface, (v) Revegetating with appropriate plant species in order to control erosion, and facilitate the development of a stable ecosystem compatible with the projected long-term use.

Amelioration methods can be custom-made for a given situation, as follows (Chadwick et al., 1987): (i) Low pH (usually < 5): Amelioration by liming. Acid-tolerant species may be planted; (ii) High pH (usually > 8): Salt content may be removed by leaching. Salt/ alkali- tolerant plants may be grown; (iii) Low nutrient status: Nitrogen deficiency may be ameliorated by nitrogenous fertilization or by growing legumes; (iv) Low moisture levels: Ridging, furrowing and mulching, etc. and growing drought-tolerant plants; (v) Soil amendment : use of other wastes, such as fly ash, slag, etc.; (vi) Planting of artificial wetlands for the treatment of acid mine drainage and polluted runoff.

Beneficial use of mining wastes

The volume of wastes generated in the process of mining increases with increased volume of mining activities, and increased mechanization. No-waste and low-waste mining technologies can in principle bring down the volume of wastes that need to be disposed of, but there is little doubt that wastes in mining cannot be avoided altogether. The use of waste rock for back-filling, recycling, and the large-scale use of wastes for the construction of roads, buildings and other civil engineering structures are some of the ways by which the wastes can be used beneficially.

Coal mining wastes **:** Taking the mining industry as a whole, there is little doubt that coal mining produces the largest volume of solid wastes. Mine gangue and coal-washing tailings are being increasingly used as filling materials, additives in concrete and for agricultural purposes.

The gangue material in coal waste tips generally has a porosity of about 35 %. The relatively high combustible content of the waste coupled with its high porosity, makes the waste liable for spontaneous combustion. It has been estimated that 40 % of the 17,000 rock waste tips in the world, are burning. Smoke from the burning tips pollutes large areas around them. An ingenious way to reduce the porosity of the waste tips and thereby reduces their proneness for spontaneous combustion, is the addition of fly ash from the wastes of the thermal power plants. In this manner, one kind of waste is made use of to reduce the environmental harm from another kind of waste!

After strengthening, the gangue material from the coal mining industry can be used in the construction of road embankments and railway lines, landscaping of building sites, and earth dams, etc. The porosity of the gangue is reduced and the strength increased by compaction with bulldozers, and addition of pore-filling materials, such as fly ash from power plants, sand, and flotation tailings. By this process, the porosity can be reduced to 20%, and the density increased to 2.1 t / m^3.

Clays with high content of organic matter can be used to make a material called karamzite. In Belgium and France, mine gangues and coal washing tailings are made use of to fabricate commercial building materials, trademarked AGRAL. The gangue material can also be made use of to make bricks, and as aggregates for light weight concrete. For instance, the brick-works of "Lvovstrojmaterialy" in Ukraine which produces 300 million bricks a year, found that the use of 10% coal wastes has reduced the consumption of fuel by 20 - 25%, besides improving the quality of bricks.

The CSIR Laboratories in India (principally, the Central Building Research Institute, Roorkee, and the Regional Research Laboratory, Bhopal) have developed innovative approaches for the use of fly ash from the coal industry and red mud wastes from the aluminium industry (vide *CSIR Rural Technologies*, 1995, p.83-88).

Clay may be mixed with fly ash (to the extent of 10 – 40 %) and made into bricks, which can then be fired in conventional Bull's kiln, or intermittent type kilns at a temperature of 950 to 1050 °C. The use of fly ash permits the production of 40 % more additional bricks from the same quantity of soil. The clay-fly ash bricks have lower bulk density, better thermal insulation and reduced dead load on the brick masonry structure. These bricks can be used for all types of construction, where normal clay bricks are used.

In areas where good quality clay is not available, fly ash-sand-lime bricks can be made. Fly ash could be used to the extent of 70 %. The bricks will have a wet compressive strength of 100 – 200 kg/ cm and water absorption of 10 to 20 %. Drying shrinkage (0.01 – 0.05 %) and thermal conductivity are comparable to those of the clay bricks. Unlike the clay bricks, the fly ash – lime – sand bricks do not need drying.

The lime – fly ash blends can be used as stabilizers in road construction. For granular soils, 3 – 6 % lime and 10 – 25 % fly ash should be used. For clayey soils, 5- 9 % lime and 10 – 25 % fly ash, need to be used.

Bricks can be made with red mud wastes from the aluminium industry. Red mud improves the quality of bricks made from clay-deficient soils. When fired, bricks made with red mud develop a pleasing pale brown, orange or golden yellow colour, depending upon the composition of the raw material, and firing temperature. They therefore have a good architectural value as facing bricks. The presence of 4 – 5 % alkalis in red mud makes for good fluxing action. Consequently, the red mud bricks have better plasticity and bonding than the normal bricks. They may be fired in the usual Bull's trench kiln.

Black coal flotation sludges can be dried to reduce their moisture content to 8 to 10 %, and the resulting product can be burnt in the thermal power plants. Brown coal sludges are finding numerous uses in agriculture. When added to the soil, the humic acids contained in coal form organo-mineral humus and sorption complexes and becomes repositories of nutrient elements. This improves the structure, pH and fertility of the soil. In Russia, the combination of manure and high-ash coal (the so-called mineral manure) proved very successful. In Hungary, brown coal dust mixed with manure is used as a fertilizer. Coal waste can be used as bio-organic mineral fertilizer.

Acid mine effluent often contains copper which can be recovered cheaply by treating the effluent with scrap iron. Methane generated in the underground mining can be collected and used to feed the boilers.

Other kinds of mining wastes: Nepheline tailings in the production of apatite concentrates can be used in the production of glass, and as a binder for silica bricks. Wastes of chalcopyrite ore concentrates can be used for the manufacture of silicate wall and facing materials, glass, etc.

Solid wastes from mining could be used as fillers in concrete and other cement-based materials. Mine soil fill material can be effectively used for the renovation of wastewater. Red mud waste is produced when bauxite is processed to produce alumina, and is available in large quantities around the bauxite mines. It contains compounds of Al (22 – 37%), Fe (24 – 26 %), Ca (2 – 4 %), Na and Si. It has been found that red mud mixed with medium-sized sand is highly effective in removing P, BOD, suspended solids and fecal coliforms from domestic sewage (Brandes et al, 1975, *Proc. Symp. National Home Sewage Disposal*, Chicago, Il., USA).

Residential and municipal wastewaters contain numerous pathogens, such as enteric viruses (which can cause meningitis and hepatitis), bacteria (which can cause typhoid fever and gastroenteritis), protozoans (which can cause amoebic dysentery and giardiasis), and helminthes (which can cause a number of chronic diseases such as anaemia and gastroenteritis). Size-wise, the enteric viruses are the smallest, and the helminths the largest. Considerations of size enter the picture because the larger the organism, the more readily it is trapped and retained when wastewater containing the pathogen percolates through the soil. Consequently, the greater the percentage of fines (silt- and clay-sized particles) in the soil, the greater is its capacity to retain bacteria. Besides, the charged nature of bacteria and viruses facilitates their adsorption on soil constituents. As it is difficult to detect viruses in soils and waste disposal systems, the abundance of faecal streptococci, and faecal coliforms are used as indicators of pathogenicity.

Excess amounts of NO_3^- may be toxic to infants and young animals, and both NO_3^- and P promote eutrophication of surface waters. NH_4^+ concentrations have decreased to background levels after percolating through 76cm of soil fill. The mine soil-fill has been found to be very efficient at removing PO_4 - P from the wastewater.

Wastewater could be applied at the rate of (say) 19.3 L /m / d on at least 0.76 m of mine soil-fill. Uniform distribution of effluent could be ensured by using low-pressure distribution or drip irrigation system (Harrison et al, 1999).

Tailings have been used in USA as bulk fill in highways, embankment material, as aggregate for sub-base and bituminous paving mixtures, in building bricks and blocks, and in the manufacture of low-grade glass (Collins and Miller, *Minerals and Environment*, 1979). China's largest gold producer, Shangdong Gold Group Co. Ltd, has recently commissioned a 4 million m^3 tailings brick manufacturing plant – it is expected to generate annual profits of Yu 12 million and pay back the company's investment in five years (*Mining Journal*, Aug. 18, 2000 issue).

Zambia converted the abandoned open pits to fish ponds.

Reuse of Mine Water

Mine water is invariably highly acidic, besides containing undesirably high quantities of toxic metals.

There is severe scarcity of drinking water in the coalfield areas of eastern India. On one hand, the water-table has gone down to 200 - 250 m due to mining activities, thus making the tapping of groundwater prohibitively expensive. On the other hand, there is abundance of mine water, which, however, is not potable because of its high acidity, and the high content of metals, such as iron . The Central Mining Research Institute, (CMRI), Dhanbad, Bihar, 826 001, India, has developed a treatment process which is claimed to render the mine water potable (item 6.2.13, *CSIR Rural Technologies*, New Delhi, India, 1995). Filtration is done adjacent to the settling pond. Two filter beds are used to work alternatively at the time of changing the bed. A slow or rapid filtration may be employed depending upon the situation. A disinfectant is incorporated in the treatment process to destroy the pathogens.

The presence of high iron content in groundwater is objectionable because of discoloration, turbidity, bad taste and tendency to form deposits in the distribution mains. The National Environmental Engineering Research Institute, Nagpur 440 020, India, developed a simple plant to remove iron from groundwater by precipitating the iron impurity as a ferric sludge (item 6.2.4, *CSIR Rural Technologies*, 1995). The plant is to be attached to a hand pump . It has a capacity of 2500 L / d (10-hr operation) and costs about Rs. 25,000 (USD 500). The plant has three chambers. "The water from the hand pump is sprayed over an oxidation chamber. The aerated water flows over baffle plates to a flocculation chamber and then to sedimentation chamber. The water then passes through plate settlers and to the filter from where the filtered water is drawn through a tap after chlorination". The ferric sludge needs to be scoured out twice a month.

Gold Processing Technology : Innovations

Gold is one metal for which there has never been a diminishing of demand. The world production of gold was about 1400 t in 1980's, and about 1800 t in 1990's. The present world production of gold is about 2500 t, worth about USD 25 billion. More countries are producing larger quantities of gold.

Where to look for gold

Gold occurs as (i) free gold ores, (ii) gold with iron sulphides, (iii) gold with arsenic and / or antimony minerals (e.g. arsenopyrite), (iv) gold tellurides, (v) gold with copper porphyries, (vi) gold lead and zinc minerals, (vii) gold with carbonaceous minerals.

The crustal abundance of gold is 5 ppb (parts per billion). Its concentration coefficient (factor of concentration of crustal abundance needed to form an economic deposit) is 2000. Mantle-derived mafic-ultramafic rocks have the highest gold contents (upto 10 ppb). In the endogenous environments, gold may be mobilized and transported in the form of thiosulphate complexes [Au (S_2O_3)] $^{3-}$ which result in coarsely crystalline gold alloyed with 50 – 75 % Ag, and chloride complexes (Au Cl_2 $^{2-}$ and $AuCl_2$ $^{-}$) which tend to be reprecipitated with concretionary Fe oxides alloyed with <0.5 wt. % of Ag. In the exogenous environments, Au may be transported as Au-humate complexes, and may end up as very fine, lateritic gold. Alluvial placers are of considerable economic importance.

Gold particles may range in size from dispersed (upto 10 mm), small (upto 0.1 mm), medium (upto 1 mm) and large (upto 5 mm). Pure gold is said to have fineness of 1000 (or 24 Karats). The generally lower fineness of gold in the greenstones (600 – 900) is attrubted to Au – Ag alloy (electrum).

Gold occurs in a large variety of environments (Hutchinson, R.W., 1987, Metallogeny of Precambrian gold deposits – space and time relationships. *Econ. Geol..*, v.82, p. 1993-2007).

Igneous **:** Basalt – ubiquitous, iron-tholeiitic; commonly found pillowed and variolitic-spheruliitic, also magnesian and komatitic; Thin fragmental-pyroclastic rock, Quartz and/ or feldspar porphyritic stock.

Sedimentary **:** Polymict conglomerate, Turbiditic greywacke, Iron formation, any facies, Carbonaceous graphitic-(pyritic) sediment.

The most important, numerous and largest major districts are of Archaen age

The environments which are characteristic of large deposits of gold, are summarized as follows (Cox and Singer, 1983, *Mineral Deposit Models*. USGS Bull., 1693, 379 p.) :

1. ***Porphyry Cu – Au*** **(Model 20 c) :** Central Cu, Au, Ag. Peripheral Mo. Peripheral Pb, Zn, Mn anomalies may be present if late sericite pyrite alteration is strong. Au (ppm): Mo (%) = 30 in ore zone. System may have a magnetic high over intrusion, surrounded by magnetic low over pyrite halo (e.g. Copper Mountain, Canada).

2. ***Hot Springs Au – Ag*** **(Model 25 a) :** Au + As + Sb + Hg + Tl higher in the system. Increasing Ag with depth. Locally NH_4, W (e.g. McLaughlin, California, USA)
3. ***Creede epithermal vein*** **(Model 25 b) :** Bleached country rock, goethite, jarosite, alunite; supergene processes often an important factor in increasing the grade of the deposit (e.g. Pachuca, Mexico).
4. ***Comstock epithermal vein*** **(Model 25 C) :** Au + As + Sb + Hg higher in the system. Also Te & W (e.g. Comstock, Nevada, USA).
5. ***Epithermal quartz – alunite Au*** **:** Au + As + Cu higher in the system. Increasing base metals with depth. Also Te & W (e.g. Iwato, Japan).
6. ***Carbonate hosted Au*** **(Model 26 a) :** Light brown to reddish brown iron oxide stained jasperoid. Au + As + Hg + W ! Mo. NH_3 important in some deposits (e.g. Carlin, Nevada, USA).
7. ***Quartz pebble conglomerate, Au – U*** **(Model 29 a) :** Braided stream channels in broad unconformity surfaces in alluvial fans. Gold gets concentrated at the base of the mature conglomerate beds deposited on an erosion surface. Anomalous radioactivity (e.g. Witwatersrand, South Africa).
8. ***Low sulphide Au – Qz veins*** **(Model 36 a) :** Arsenic best pathfinder. Association with Ag, Pb, Zn and Cu. Abundant quartz chips in the soil. Gold may be recovered from the soil by panning (e.g. Ballarat Goldfield, Australia).
9. ***Homestake Au*** **(Model 36 b) :** Volcanogenic gold, iron formation – hosted gold. Archaean lode gold. Au + Fe + As + B + Sb (+ PGE in mafic volcanic terrains). Bi , Hg and minor Cu – Pb – Zn – Ag – Mo (e.g. Vubachikwe, Zimbabwe).
10. ***Placer Au – PGE*** **(Model 39 a) :** Anomalously high amounts of Ag, As, Hg, Sb, Cu, Fe and S. Heavy minerals, magnetite, chromite, ilmenite, haematite, pyrite, zircon, garnet and rutile. Au nuggets have decreasing Ag content with distance from the source.

How to look for gold

Since pyrite is a precursor for mineralisation of metals, such as gold, it follows that the higher the content of pyrite in the host rock, the greater the possibility of gold mineralisation. A 1:1 correlation has been found between the pyrite content of the host rock and its Au content .

M◊ssbauer spectroscopy and Raman spectroscopy have emerged as useful tools in looking for gold (Ferrow, 2001). M◊ssbauer spectroscopy (MES) is useful in determining the valence state, coordination number, spin state, magnetic properties and structure of the minerals. For instance, the MES spectra of Au-poor pyrite samples are characterized by low-spin doublet, while Au-rich samples contain an additional magnetic sextets (produced by the oxidation of pyrite substrate during the simultaneous reduction and sorption of Au.) (Fig. 15.4; source: Zhenru et al, 1989). The Raman spectrum of Au-rich quartz is markedly different from the spectrum for Au-poor quartz – the intensity of Raman emission for Si-O-Si and for crystal lattice vibrations is higher for Au-rich quartz than for Au-poor quartz.

The most serious problem that hindered the prospecting for gold has been the great difficulty and expense in determining the gold content *in situ* at ppm and sub-ppm levels. The usual practice has been to pan for gold (which requires water) and look for grains of gold visually. Fire assay of gold is accurate, but it cannot easily be done in the field.

Portable XRF devices are available for the geologist to check the ore grade in the drill core, or this could be done automatically. Niton's new XL-500 Prospector can assay ore samples directly *in situ* (rock face or drill core). It is a single-piece, hand-held analyzer weighing only one kg., including the battery. Typical *in situ* measurements range from 30 – 60 seconds, and 500 – 1000 measurements can be made per day. About 1000 measurements can be stored in the instrument internally, and can be downloaded as needed for mapping, grade control and other kinds of analyses. Niton also markets a special device for precious metals (called Precious Metals Analyser), for the analysis of Au, Ag. Pt, Rh, Ru, Ir, Pd, Cu, Zn, Ni, Co and Fe in ores, and fire assay can be avoided. Details about Niton instruments can be had from www.niton.com.

How to extract gold

McNulty (2001, Cyanide substitutes, *Mining Mag.*, v. 184(5), p. 256-261)) gave an excellent update on cyanidation.

The cyanidation process was patented in U.K. on October 19, 1887 by J.S. MacArthur and two brothers, W. and R.W. Forrest. Cyanidation changed for ever the economics of gold industry. For instance, the application of cyanidation process in the Rand goldfields of South Africa, led to a thousand-fold increase in gold production in a matter of three years – from 300 oz (9.33 kg) in 1890 to 300,000 oz. (9331 kg) in 1893. During the past 20 years, cyanidation accounted for about 92 % of the total world production of gold. Cyanidation has the following advantages: (i) it requires only dilute solutions containing typically 300 – 1000 ppm (0.3 – 1.0 g / L) of sodium cyanide, (ii) the pH range used (9.5 – 11.5) is such that only gold and silver get mobilized, and (iii) it is simple to operate and control.

Heap cyanidation of low grade ores has proved to be efficient and inexpensive, and is extensively used all over the world. The great advantage of cyanide heap leaching is that there would be no discharging of process solutions, and minimum recycling of water. Treatment and discharge of process solutions would not be needed during the operation. In effect, there would be a single permanent large heap leach pad. Percolation of pregnant cyanide solutions downwards through hundreds of metres of leached ore can take place, without the solutions undergoing chemical change. On the other hand, other lixiviants (such as, sodium hypochlorite stabilized by sodium chloride, Bromine stabilized by sodium bromide, Ammonium thiosulphate stabilized by ammonia, and catalyzed by cupric ion, and Acidic thiourea) require rigorous control of pH and Eh, and there is always the possibility of side reactions and precipitation of gold.

Cyanidation may be used to extract gold from almost any kind of gold ore. It may be done in different ways : (i) leaching of the run-of-mine or crushed ore, (ii) vat leaching, and (iii) Leaching of ground ore, flotation concentrate, etc. in agitated tanks. Activated carbon is being increasingly used to recover gold dissolved by cyanide in ore pulps (carbon-in-pulp process) or in clear pregnant solutions (carbon-in-column process).

The methods of treatment applicable for different kinds of gold associations are summarized as follows (Weiss, 1985, *Mineral Processing Handbook*, SME, New York) :

Alluvial gold : 1. Gravity concentration, 2. Amalgamation

Free milling lode ores : 1. Gravity concentration, 2. Amalgamation, 3. Direct cyanidation, activated carbon in pulp.

Free milling sedimentary ores : 1. Direct cyanidation, 2. Treatment of refractory carbon, direct cyanidation.

Gold tellurides : 1. Bulk flotation-roasting cyanidation, 2. Direct cyanidation – SO_2 roasting of concentrate cyanidation, 3. Flotation – cyanidation of concentrate-roasting of residue – recyaniding, 4. Direct cyanidation, with added bromocyanide.

Gold with pyrite and marcasite : 1. Flotation – smelting of concentrates, 2. Flotation – cyanidation of concentrates.

Gold with pyrrhotite : 1. Direct cyanidation with pre-aeration at low lime alkalinity, 2. Direct cyanidation – flotation of cyanide tailings – regrind and recyanide flotation concentrate or roast and recyanide.

Gold with arsenopyrite : 1. Direct cyanidation, 2. Flotation – roasting of concentrates, 3. Roasting ore – washing – cyanidation, 4. Autoclaving, 5. Nitric acid oxidation.

Gold with copper ores : 1. Flotation – smelting of concentrates – recovery during electrolytic refining, (2) Flotation – cyanidation of molybdenum.

Gold in refractory : 1. Roasting – cyanidation carbonaceous ores, 2. Chlorination of ore – cyanidation, (3) Flotation of graphitic material – cyanidation of tailings.

Gold with lead-zinc ores : 1. Flotation – smelting of concentrates, 2. Jigging – amalgamation – retorting.

Degussa-H♦ls of Germany has developed proprietary technologies which has the effect of making the gold recovery more efficient, while at the same time ensuring minimum or nil adverse impact on the environment:

PAL – Peroxide-assisted leach to increase gold recovery,

CCS – Cyanide control system to optimize cyanide consumption,

DETOX – Cyanide detoxification technology to meet the environmental standards.

From the environmental point of view, the peroxide-based detoxification technology is most relevant. The detoxification technology can be custom-made for any kind of mining effluent.

Author Index

G

H

I

J

K

L

M

N

O

P

R

S

Subject Index

A

C

D

E

F

G

S

W

www.ingramcontent.com/pod-product-compliance
Lightning Source LLC
LaVergne TN
LVHW022047250826
846425LV00029B/16

* 9 7 8 9 3 5 2 3 0 0 2 2 8 *